Into the Sciences

by Frederick J. Ross

Into the Sciences

ISBN 978-1517629892

For Jim Tarvid

Contents

1 | Introduction

When I was an undergraduate student studying physics at university, my classes presented cleaned up theories. The messy details of experiments were not mentioned. How the mathematics we were using related to reality was rarely touched on. The scaffolding used to erect the theory had been carefully removed. The history of the field's development was reduced to a few large leaps, and those leaps were assigned to the credit of a few famous names.

So we left these classes facile in manipulating formulas, but mystified by how you would actually measure the energy of the single bound state of the helium molecule ion. We could calculate the Grüneisen ratio in thermodynamics, but had no idea why we would care about it. We learned that Einstein formulated special relativity, but not that Poincaré, Lorentz, and a previous generation of workers had derived the mathematics that Einstein used by exploring the consequences of Maxwell's theory of electromagnetism.

This is understandable. A lecturer in physics or sociology is faced with passing on some degree of technical competence in limited time. The expectation is that she will try to save time by cleaning up the mess, and that she is not teaching a history course.

But the orderliness of the structure implies that the work is basically finished, nor does the student get to see how a subject develops. For those who will only use the material as a tool, this is plenty. A structural engineer professionally need not care what led to Newton's mechanics. For a physicist, who is supposed to construct such material rather than use it, this hides the primary models that underlie what she is being trained to do.

Worse, the simplification of a field's history from all the actions by many individuals to a few names and steps has a disturbing subtext: unless you're one of the chosen, the blessed, the brilliant, any work you do is just filling time until the next great mind comes along. A few practitioners, convinced that they are one of the great men, will even openly espouse this view, as David Fischer does in *Historians' Fallacies*:[23]

> [A] general interpretation is fashioned by an essayist not as a heuristic hypothesis but as an affirmative proposition. In the next twenty years or so, a legion of gradgrinds manufacture monographs which reify the essay, with a few inconsequential changes...This process continues until another essayist publishes another brilliant general interpretation, and another generation of gradgrinds are wound up like mechanical rabbits and set to running about in ever-smaller circles.

What a horrid, damaging narrative to inflict on the student or young practitioner. For most students that narrative is softened somewhat by working under the guidance of a skilled practitioner, but it never quite goes away.

Nor is it only a practitioner's relationship with her own field that can be toxic. I remember making the acquaintance of a toxicologist while waiting for a ferry. When I explained that I was trained as a physicist, her comment was, "Oh, real science." Similarly, physicists have collections of jokes about biologists and social scientists. From physics and chemistry to history and musicology, these are all "real sciences." But if a practitioner's understanding of what is scientific was acquired by osmosis from one or two practitioners, her view of science will necessarily be parochial, and the practices of other fields will seem alien.

That alienness keeps people within the boundaries of their fields. I was originally trained as a physicist, then forced by circumstance to retrain as a biologist, in complete isolation from physics. It took a year before I started to absorb the different ways of thinking, and during that year I regularly reacted with alarm as some process of thought apparently made no sense. I was in graduate school at an institution that was solely biology, and there were a number of physicists there who had already made the transition who talked me through some of the mental difficulties or I would likely have retreated.

That transition, earlier exposure to musicology, and later dabbling in history and the social sciences forced me to reformulate an idea of science for myself that described all of what I was seeing. This book aims to set down that result, which provides the young practitioner with a view of science that is both more accurate and more empowering than the narrative of great men. It is meant to describe all the sciences, from chemistry to musicology, not just certain fields chosen as ideals. It replaces the amorphous activity of "research" with concrete activities and modes of progress. And it provides a framework for a practitioner to understand why other fields than her own exist, why they don't seem to operate the way hers does, and how to efficiently go about learning another field if she feels so inclined. The examples are unfortunately biased towards the physical and biological sciences, since that is where I have the most experience, but, despite that, I hope at the end you will know how to delve into the messy, unsanitized realities of a field and emerge with a useful understanding of how it works.

Lastly, a quick note on words: I eschew the word "scientist," which carries too much baggage. Our society loads the word with images of lab coats, professorships, lonely geniuses, and men. The truth is very different. Some wear lab coats. Many don't. Most aren't professors. They almost all work in collaboration. And they are mostly women. Thus I refer to someone who does science as a "practitioner," and I use the female pronoun to describe that person.

2   Idealized trials

Let us begin with the activities that occupy the vast majority of practitioners' time: gathering data and analyzing it. It could be an astrophysicist measuring the spectrum of the light emitted by a celestial object; a microbiologist measuring how fast a population of bacteria in a liquid culture increases; a botanist producing illustrations of plants for a flower atlas; an agronomist measuring the effect of a fertilizing regime on a crop.

I will refer to all of these activities as "trials." It is important to note that many of these are not experiments. They are observations of phenomena outside the control of the practitioner. In these cases, practitioners seek "naturally occuring experiments," situations close to how the practitioner would have designed the experiment were it possible, and there are whole fields that have no experiments whatsoever. Astrophysics must work with the stars and galaxies that exist, and history has no way to get a second historical record.

Having only observations is not an insurmountable limitation. The strongest evidence linking smoking and lung cancer comes, not from experiments, but from long-term observation of identical twins in Scandanavia.[25, 42]

Nor should we artificially limit the notion of a scientific trial. A botanist producing illustrations for a flower atlas is performing a trial, though its outcome—a drawing—is very different from what is produced when a physicist measures the speed of light, and different again from the trial of an historian producing an account of the past.

But we see children pouring liquids back and forth or closely watching bugs. What makes a chemist testing for the presence of mercury a scientific trial and a child pouring liquids back and forth not? The different is one of intent. The chemist has in mind some idealized trial which she is trying to approximate. The idealized trial is conceived so that, if it were perfectly achieved, it would impart knowledge about the world. For example, all of the following, if no details trip up the practitioner, will produce a solid piece of knowledge:

- Weigh the amount of corn produced by two strains grown next to each other to find which one yields more.
- Compare the number of objects of north African origin that are found in Gaul that date to before and after the Goths replaced government by Rome in that region to see how the volume of long distance trade changed.
- Hook up a piece of material to a pair of electrodes and measure the resistance across the material electrodes while slowly cooling it until the resistance drops to zero to find the superconducting transition temperature of the material.
- Select a number of specimens of dandelion and use them to make a drawing capturing the important features that identify a dandelion.

There is no universal logic of idealized trials. Those readers hoping for an onslaught of symbolic logic or hard rules about the form of idealized trials will be disappointed. Idealized trials aren't derived from pre-existing logic. Rather, the logic is developed to describe existing trials. As new idealized trials are developed, the logic is expanded to account for them.

On the other hand, certain constraints were developed in the 17th century that we now consider essential for an idealized trial, the most important of which is the expectation that an independent practitioner can reproduce the results of the trial. Reproducibility is a negative criterion. It says, "You cannot use this idealized trial." and "These trials do not agree. At least one of them cannot be accepted."

Making a trial "reproducible" is harder and more subtle than it would first seem. What does it mean to reproduce results in a field like history where the historical record is fixed? If we present two historians with the same documents and subject, we cannot expect word for word identical accounts. If one of the historians is removed in time from the other, we would have an enactment of Borges's story *Pierre Menard, Author of the Quixote*[10], where an author labors to make himself such a creature that, in the 20th century, he will produce an original work that is word for word the same as *Don Quixote*. Even in experimental sciences, measurements are messy. If one physicist measures the Young's modulus of copper to be 117.3Gpa, and another measures it to be 117.41GPa, do we say the result is not reproducible? It is actually quite reproducible, but we cannot expect lockstep.

We might try to argue our way out of this latter situation by saying that the physicists should report a value with a 95% confidence interval, but, in practice, this rarely works due to Hamming's rule: "90% of the time the next independent measurement will fall outside the previous 90% confidence limits." Why? As Hamming puts it,

> Consider how you, in fact as opposed to theory, do an experiment. You assemble the equipment and turn it on, and of course the equipment does not function properly. So you spend some time, often weeks, getting it to run properly. Now you are ready to gather data, but first you fine tune the equipment. How? By adjusting it so you get consistent runs! In simple words, you adjust for low variance; what else

can you do? But it is this low variance data you turn over to the statistician and is used to estimate the variability.[33]

We can see this in practice in particle physics. Consider the measurement over time of the mass of the η particle:[4]

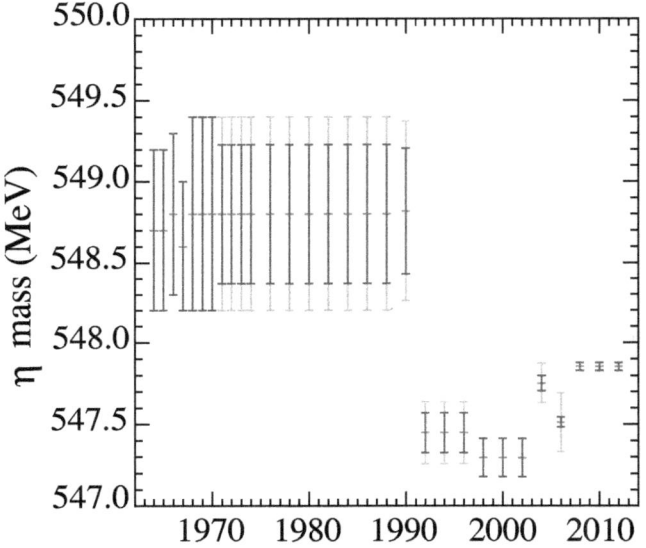

Each vertical bar represents a measurement of the mass of this fundamental particle. The lower end of the bar is a lower limit on the mass, the upper end an upper limit. The bars are arranged chronologically. But the bars do not all overlap vertically, meaning that the experiments disagree. In the early 1990's there was a jump in both precision (shown by the shorter bar) and in the measured value, followed by more shifts and more increases in precision. This would seem to indicate that the trial is not reproducible. But when we examine the actual numbers we find that all of the trials give values between 547.0 MeV[1] and 549.5 MeV, a variation of less than half a percent, which seems quite repoducible, despite clear evidence of Hamming's rule. And worries about error bars and the like barely apply in anthropology or history.

Arguments about reproducibility have been going on since the rise of the modern sciences,[63] and despite its seeming difficulty, we have made progress. Looking back, seventeenth century practitioners seem almost fussy to modern counterparts in their obsessive noting of detail in hopes of reproducibility.[27]

I think the clearest example of why anyone cares about reproducibility comes from musicology. Performance practice in western music—the conventions of how to interpret a piece of written music—has traditionally wandered from generation to generation. Each generation only had the performance of the previous generation from whom they had learned as a basis, so it could wander without limit. Over the course of several hundred years since the seventeenth century composer Archangelo Corelli wrote his music, the performance practice drifted to one of slow tempos, no dynamic variation, no improvisation or decoration, a practice that produced a lot of performances best described as insipid. We know today that this is not at all how Corelli played.

Nor does it take hundreds of years for performance practice to dramatically change. Sergei Rachmaninoff wrote in the late nineteenth century, only two or three generations of performer ago. Today his works are usually played with lots of rubato and schmaltz. Yet we know he did not play that way. He played his works at one point on a player piano, which precisely recorded his performance on rolls of punched paper.[2] When we play those

---

[1] An MeV is a mega electron volt, a convenient and ubiquitous unit in particle physics.

[2] The player piano is most familiar today as the piano in the corner of

same rolls back in a player piano today, we hear renditions that are precise and rhythmic.

We have no such access to Corelli or to any musician before the mid nineteenth century, but we do have the instruments that were played and many treatises written at the time. Starting in the early 20th century, a new field arose called musicology that tried to infer a performance practice from these artifacts in a reproducible way (with many of the same problems as the field of history). In the process, the field reclaimed several golden ages of music for us which had otherwise been almost lost, such as the viol consort music of England in the 17th century.

A performer today can work with a performance practice for Corelli that has a clear justification as a starting point. Of course some areas remain unknown, such as the exact nature of the tuning that J.S. Bach wrote his Well Tempered Clavier to show off. We have no way of knowing it, though there are many proposals, most of which are reasonable given the documentation available, and many of which sound utterly wild to modern ears. But despite these holes, we have a clear starting point for most western music since the twelfth century. A performer may depart from this practice, and often does, but she can start with a practice that she can defend as reasonable and work from there.

Let us turn now to the progress of idealized trials. New idealized trials open up new possibilities for what can be studied. For example, we cannot study nature versus nurture in humans the way we do in mice. In mice, we would perform controlled breeding and raise the offspring in different environments to see which aspects of their behavior are due to heredity and which are due to environment. In humans, this is unethical. Our society does not allow scientists to force people to breed, nor to decide the fates of children, and, even if it did, such a study

---

saloons in western films that's playing by itself.

would take decades to run as opposed to a few years in mice. The solution to this problem is to use naturally existing breeding experiments, notably twins. In a population of twins, all the differences measured in identical twin pairs should be due to environment, while in fraternal twin pairs it will be due to both environment and heredity. The difference between fraternal and identical twins wasn't demonstrated until the 1920's,[56] but as soon as it was, twin studies became a mainstay of epidemiology and human genetics.

Existing idealized trials also shift over time. Daston and Galison open their book *Objectivity*[14] with a particularly clear example of this shift, taken from the history of fluid mechanics. Arthur Worthingon, from 1875 to 1894, painstakingly studied what occurs when a falling droplet hits a surface by illuminating his darkened laboratory with a precisely timed, millisecond flash, then sketching the latent image impressed on his retina. From thousands of such sketches he assembled a description of the behavior of splashing droplets, describing various outcomes, all characterized by the utter symmetry of the splashes.

In 1894, he first succeeded in capturing the image on a photographic plate instead of his retina, and the illusion of symmetry vanished. Images that had seemed symmetric seen with his eye were revealed to be irregular on the photographic plate. Worthington could no longer believe what his eye told him about the symmetry of a form recorded by a latent image, and photographic plates replaced the eye.

Such large scale shifts do not replace everything that came before. In history, for example, objectivity—removing the human from the production of information—is meaningless. An historian trying for objectivity, for the purely mechanical selection of her facts, is engaging in what Feynman called "cargo cult science"[3]:

---

[3]The reality of cargo cults is actually a fascinating piece of anthropology,

> In the South Seas there is a cargo cult of people. During the war they saw airplanes land with lots of good materials, and they want the same thing to happen now. So they've arranged to imitate things like runways, to put fires along the sides of the runways, to make a wooden hut for a man to sit in, with two wooden pieces on his head like headphones and bars of bamboo sticking out like antennas—he's the controller—and they wait for the airplanes to land. They're doing everything right. The form is perfect. It looks exactly the way it looked before. But it doesn't work. No airplanes land. So I call these things cargo cult science, because they follow all the apparent precepts and forms of scientific investigation, but they're missing something essential, because the planes don't land.[20]

Another shift in idealized trials occurred in microbiology and molecular biology starting in the late 1990's. Microbiological work from the 1970's to the 1990's focused on properties of microbes as though they were the same everywhere. For example, when the amount of RNA transcribed from a locus of DNA changed due to a change in the microbe's environment, microbiologists thought of it as a fixed change in the number of molecules, the same in all cells of the population, and starting from and arriving at fixed levels.

A series of papers published in the first decade of this century[19, 65] by a number of physicists—whose training told them that any system on the scale of chemical reactions in a cell would be noisy and variable—changed that. The strange part of this tale is that there was a literature from the 1920's on

---

and their effectiveness in Feynman's parable is more a commentary on our own culture than on the cults in question.[52]

just such variation, exquisitely documented and illustrated, and completely forgotten.[36]

To summarize, the basic activity of practitioners is running trials driven by trying to match some idealized. We have also seen our first two modes of progress: gathering data from trials, and creating and refining idealized trials. Next we look at how practitioners judge whether a trial has reached its ideal.

3 Reasonable approximations

We now turn from idealized to real trials. When running a real trial, we find ourselves in slippery, messy territory. Any real trial is at best an approximation of an idealized trial. To return to an example from the previous chapter, when we grow two strains of corn and measure their yield, we have to account for lots of other details. Which varieties of corn will get planted where for the trial? Are some corners of the field wetter or dryer? Are they cross pollinating? What watering and fertilizing schedule should we use? Would our results differ if we used a different fertilizer?

How do we know that we have accounted for everything we must account for? Quite simply, we don't, and trying to reason our way to a perfect approximation of an idealized trial is paralyzing and futile. Given a full list of the possible errors that we might commit, we might reason our way to a perfect approximation, but all we will produce in practice is a list of known possible errors. This is clearer in hindsight. Consider the first recognizable clinical trial in the west, done in 1747 by James Lind of the British Navy.[5] He took eight sailors suffering from scurvy, and tried four treatments on them in parallel, each treatment given to two men.

Lind did not allocate the treatments randomly. Instead he gave what physicians at the time suspected was the most effective remedy to the patients in the worst shape. Today we would insist on randomizing which patients got what treatment, but expecting it in the eighteenth century is unreasonable. It took the development of probability and statistics during the nineteenth century before the randomized trial appeared in 1883[31, 54] Lind did not even have the vocabulary to speak of randomization. It is worth keeping in mind that we are certainly as blind to the possibilities of error in our own practice.

Interestingly, scurvy trials have certain features that make them less susceptible to the errors that our current practice tries to avoid. Sailors long at sea did not spontaneously recover from scurvy, which removed many of the sources of error that randomization tries to prevent, and the remedies that Lind used as an eighteenth century physician were almost all placebos, whether he knew them to be or not, so he was effectively blinded and using placebo controls. By the nineteenth, century the British Navy, using only this degree of sophistication in their clinical trials, had reduced deaths aboard ship due to disease by two thirds.[7]

Returning to our argument, if we cannot, even in principle, perfectly approximate an idealized trial, how are we to say if a given approximation is adequate? The best answer I have found to date is the "reasonable person" standard from law. Its first articulation in 1837 is as clear as any:

> The care taken by a prudent man has always been the rule laid down; and as to the supposed difficulty of applying it, a jury has always been able to say, whether, taking that rule as their guide, there has been negligence on the occasion in question. Instead, therefore, of saying that the liability for negligence should be co-extensive with the judgment of each individual, which would be as variable as the length of the foot of each individual, we ought rather to adhere to the rule which requires in all cases a regard to caution such as a man of ordinary prudence would observe. That was, in substance, the criterion

presented to the jury in this case and, therefore, the present rule must be discharged.[69]

Thus Lind's scurvy trial was reasonable for his time, but not today. And reasonable practice varies not only with time but with field. The first randomized clinical trial took place in 1946,[5] decades after randomization first appeared in Peirce's work in psychology in the 1880's.[54]

The details of how a practitioner approximates an idealized trial, and, in esoteric fields like particle physics, what idealized trial is being approximated, are too often overlooked when teaching or popularizing science. Yet these details constitute the majority of the knowledge of a field. Is it obvious that you should paddle your canoe upwind while collecting microbe samples from a freshwater lake? Or that part of setting up your microscope for use is removing the eyepiece and adjusting the image of the lamp seen through the eyepiece tube until it is sharp before focusing on the specimen? Without these details there is no reasonable practice for practitioners to share.

An example of how the reasonable person standard is refined and tightened comes from genetic studies of tuberculosis. A common experiment in tuberculosis laboratories is to produce mutants of the bacterium, and infect one group of mice with the mutant and one group with its unmodified ancestor, then measure whether the mutant kills the mice more or less quickly than its ancestor, or attains a different population size in the lungs of the mice than its ancestor. With a careful choice of mutants and of strains of mice, these experiments reveal factors that affect how tuberculosis infects its host.

My coworker Anna Tischler's work is a particularly nice illustration. She was looking at tuberculosis's mechanisms for countering the immune system, in particular the mechanisms triggered by a small molecule called interferon-γ that is produced by a class of white blood cells called T cells when they are exposed to tuberculosis. Interferon-γ triggers chemical attacks by another class of white blood cells called macrophages. Anna found a mutant of tuberculosis that survived better than its ancestor in mutant mice that did not produce interferon-γ, but did not survive better in strains of mice that did produce interferon-γ, but lacked the known mechanisms that interferon-γ triggered. This showed that there was another, unknown mechanism that was triggered by interferon-γ, and that tuberculosis had a specific response to it.[66]

Once you find such a mutant strain, to be certain that the effect that you see is due to the mutation you know about and not some other, random change, you must add a copy of the original form of the locus to the mutant and see if the effect goes away. This process is called "complementing the locus."

Megan Kirksey, another researcher in a lab where I was working, suffered from a very clear example of why complementing is part of reasonable practice. She discovered that the time growing in liquid cultures needed to make the mutant strains of bacteria caused another, spontaneous mutation in many of her strains.[43] These mutants did not produce a small, fatty acid usually found on the surface of the bacterium, which let the mutants grow slightly faster in the laboratory than their ancestor and out compete it, but which made them less effective at infecting animals. When the mutants and their ancestors were tested in mice, many of the intentional mutations appeared to have an effect on the virulence of the strains, when the effect was really due to the absence of this small fatty acid.

My colleague discovered this problem because she tried to complement her mutants, and found that restoring the loci that she had disrupted did not affect the behavior in mice of any of her mutant bacteria. Although it is common for some mutants not to complement, in this case it occurred in too many for that to be plausible.

It turns out the particular strain of tuberculosis our laboratory and many others were using had this tendency. In short order, the field switched to a different strain which did not readily mutate to stop making the small fatty acid in question. However, this change meant that the laboratories had to recreate their mutant libraries, and we still don't know how many results in earlier papers are affected by to this problem.

Another example of how the reasonable person standard evolved can be seen in the field of Norse history. Traditionally, bodies found in Norse burials were classified as male or female based on the grave goods surrounding them. Bodies found wearing armor and with weapons in their hands were classified as men. Bodies with spinning tools or other domestic implements were classified as women. However, when pathologists with no information about the grave goods inspected the skeletons they found that this was untenable.[50] The chances of a pathologist mis-sexing a well preserved skeleton are very low, and women bearing weapons and men bearing distaffs were common.

The reasonable standard can be different in different communities. This is best illustrated by practitioners working on the edges between fields. Albert-László Barabasi, who studies applications of scale invariant networks in biological systems, is considered an inspiring and inspired practitioner among a number of biologists who lack the training to follow his mathematics. Unfortunately, his methods fail to meet the reasonable practice of statisticians and mathematically trained biologists, and all of his claims have been shown not to hold when his mathematical and statistical errors were corrected.[53]

The progress of the reasonable person standard is sometimes very slow, only occurring in some cases with the death of older practitioners. The study on mis-sexing bodies in Norse burials referred to above was already two years old when archaeologists studying a new and magnificent Etruscan burial in Italy[48] proceeded to (incorrectly) guess the sexes of bodies based on the presence of weapons or domestic implements.[71] Yet this was a society well known even in antiquity for the independence and strength of its women.

A similar systematic error was pervasive in the social sciences until the latter half of the twentieth century, that of ignoring the existence of women or indeed of anyone who was not a privileged elite. Stated so baldly, it seems unreasonable. But the intellectual inertia related just to ignoring women was so strong that a different department, women's studies, had to be formed in order for the remedial work to be carried out in universities.

The history of science was similarly afflicted until the 1970's. Historians of science were writing accounts of how the predominant theories of their day came to be. A group at the University of Edinburgh pointed out that this was horribly biased. They proposed the Strong program, where all the approaches to a phenomenon that were being worked on at the same time need to be treated without privileging whichever one gained consensus afterwards.[8] The whole historical record pertaining to the study of some phenomenon must be examined not just the documents on the path leading to the present consensus.

Similarly, history also has a tradition of "great narratives", such as those of Toynbee[67] and Durant[16]. Today, both of these monumental works are considered unreasonable because of the manner in which they selected their facts for their world histories. For example, discuss the thousand or so years of Homeric and classical Greece and western Rome, but barely mention Persia or China, and do not touch on Aksum or the Olmec at all. For what is supposed to be a world history, it is hard today to imagine any reasonable criterion of selection that would result in these choices. If the criterion of selection was complexity of material culture, all of these cultures would call for fairly equal treatment. If the criterion of selection was social stratification and complexity, then later in their world histories we would expect to see pre-European contact Tahiti discussed alongside France, which we notably do not.

The only criteria of selection which can lead to this choice are the availability of information to the practitioner or choosing areas to focus on based on the perception of them leading to the polities to be discussed later in the work. The first of these is pragmatic, but not really defensible. The second is the same fallacy which the Strong program in the history of science was meant to counter.

All these examples illustrate the tightening of reasonable practice in the past. Is there any way to tell how well contemporary practice works? For a limitd range of fields parapsychology acts as such a check. After centuries of study, during which any positive result was later found to be due to chance or methodological error, we can reasonably assume that the phenomena it studies don't exist. This means that all positive results in parapsychology can be taken as indicators of errors. Since parapsychologists are very careful methodologists, and share much of their technique of blinding, randomization, and experimental design with the biomedical sciences, the field lets us study how well the reasonable practice of contemporary biomedical science actually works.

Let us turn from the evolution of reasonable practice to what that evolution does to a field. From the examples above, obviously many results are invalidated. But not all old trials are list. We have not learned anything which invalidates Koch's trials to determine the etiology of tuberculosis.[44] Nor does a particular trial being invalidated bring a scientific structure crashing down. Some of Snow's trials on cholera have been invalidated, but he tackled the problem from so many directions that those few may be discarded without materially altering our understanding.

The reasonable person standard depends not only on when a

trial was done, but in what field. A biologist studying the microbiology of freshwater lakes will take pains to document that she paddled her canoe upwind while collecting to avoid contamination from her boat. Such a precaution would not occur to a practitioner trained in another field, even one as close as medical microbiology.

Yet there are aspects of reasonable practice that are common to a range of fields: designing experiments with controls and randomization, plotting data, calculating averages, fits, confidence intervals, or p-values. These are taught under the name of statistics. But even what statistics are taken as reasonable at a point in time depends on the field. Each field's practitioners tend to carve off a particular part of statistical practice and perpetuate it as part of their reasonable person standard.[1] Particle physicists use a small palette of statistical methods, mostly least squares fitting of curves and χ^2 tests of the quality of those fits, techniques which date to the late 19th and early 20th century. Statistical practice in laboratory biology and agronomy was carved off in the mid 20th century, largely by Ronald Fisher[24], though it is often corrupted in practice. Molecular biologists and clinical researchers, in particular, are notorious for relying on doing trials three times, or on three patients, which has no justifiable basis. Similarly, some backwards fields of biology to this day reject all innovations in displaying data since the invention of the basic scatterplot in the nineteenth century.

The most peculiar statistical offshoot is machine learning in artificial intelligence. Machine learning began by carving off a piece of the formal structure of statistical inference to experiment with building machines that could develop behaviors based on examples of stimulus and expected response. Machine learn-

---

[1] I would love to see a good history of the creation and perpetuation of such statistical sublanguages.

ing was a dramatic success. Essentially all mail in the United States has been sorted by computers for decades, and the best efforts in the field match a human's ability to identify objects or faces in pictures. Yet it has also reinvented, or worked around the lack of, much material from statistics that was not originally carved off.

Statistics constitutes a small part of the practice that makes up the reasonable person standard for a field. A molecular biologist spends several years learning how to handle liquids in a sterile manner so as not to disrupt the enzymes involved, and to avoid all the small, unarticulated errors that can cause experiments in that finicky field to go wrong. Subfields have even more specific practice. A molecular biologist starting to study RNA in a deep way can expect to spend months learning how to prevent contamination of their trials by the ubiquitous, RNA destroying enzymes which all naturally occurring cellular organisms produce as protection against viruses. Until the 1980's, chemists and biologists learned to blow glass to create apparatus. Physicists still learn to build electronics, machine metal, and work with liquid handling and vacuum systems. By comparison, most molecular biologists and physicists spend at most a few tens of hours learning material from the shared body called statistics.

I want to emphasize once more that distinct fields have different reasonable practice. Those differences regularly lead practitioners astray when looking at fields besides their own.

4 | Primitive notions

So far we have discussed trials. But what are the trials trying to produce? I have been vague and said "knowledge," but it is time to be more precise.

Practitioners end up measuring the same thing many different ways. Particle physicists have many different ways of measuring momentum. Biologists have many ways of measuring the growth of microbial populations. We see this again and again across fields. Many different trials can produce values of the same quantity. This isn't a coincidence. Most of these trials were *designed* for this. We call the thing these trials are designed to produce a value of a "primitive notion."

Primitive notions take their name from their use in building theories to summarize trials. The trial produces a value of a primitive notion and, if we accept the trial as reasonable, we use that fact that the primitive notion has that value as an assumption in our theories. The result of the trial is treated as a primitive, an assumption or axiom, in the theory.

Primitive notions, like reasonable practice, must be shared by a community to be useful. When driving a car, we all agree on the primitive notion of speed, and know a plethora of rules concerning it. People living in industrialized societies all agree on a primitive notion of "on time," which is quite foreign to many other societies. On the other hand, achievement tests in schools are a source of continuous argument and acrimony because no on agrees on a primitive notion that the tests are supposed to measure.

The example of achievement tests highlights a crucial point: a primitive notion needs to be defined in terms of trials, which is known as operationalizing the primitive notion. A surprising amount of work goes awry at this point. The problems that come up are most articulately described by the "Unlabel Me" movement in applied behavior analysis, which arose as a reaction against using descriptions such as "aggressive" or "agitated" in describing and working to alter behavior. "Agitated" is a meaningless label, as Susan Friedman demonstrated:

> I asked the students in one of my parrot behavior classes to list three behaviors a parrot would display if it was labeled an easily "agitated" parrot. As predicted, they submitted twenty different behaviors (bites, paces, screams, etc.) but the really telling piece of data is that only 9 of the twenty behaviors appeared on more than one person's list! To the extent that we remove ourselves from describing observable, measurable behaviors, we reduce our ability to understand, predict and change behavior.[26]

The same lesson arose in quantum mechanics in the early twentieth century. What is now referred to as the "old quantum theory," based on Niels Bohr's notion of fixed, discrete orbits for electrons, "may be seriously criticized on the grounds that [it contains], as a basic element, relationships between quantities that are apparently unobservable in principle, e.g., position and period of revolution of the electron."[35]. Heisenberg's solution, presented later in the same paper, is "to try to establish a theoretical quantum mechanics, analogous to classical mechanics, but in which only relations between observable quantities can occur."

The lesson has not been universally learned, however. The canonical example of failing to work in terms of operationalized primitive notions is the sizable industry of personality profiles. All these profiles, such as the Enneagram, the Myers-Briggs Type Index, and the Hermann Brain Dominance Index assign people to one of a set of categories. Those categories, however, were never defined operationally. In the Enneagram,[38] they were created by Oscar Ichazo based on a Sufi figure popularized by the mystic G.I. Gurdjieff. The Myers-Briggs types[13] were made up by a mother and daughter from their reading of Carl Jung. The Hermann Brain Dominance Index was put together by Ned Hermann of General Electric,[37] based on oversimplified or outright false notions of brain function.

The questionnaires used to determine an individual's "type" in any of these systems are created by choosing a set of categories, selecting a group of people to study, and assigning them each to a category at the practitioner's whim. Then each member of the group is given a long list of questions to answer. If the answers to a question correlate well with the category individuals are assigned to, the question is kept. The rest are discarded. Then six months or a year later, those questions are given to the members of the group again, and those that still correlated with the category are kept for the final questionnaire.

The companies pushing these profiles go through lots of statistical contortions to demonstrate the validity of their questionnaires, but never address the key problem that the categories were chosen by individual whim. There is no reason to believe that they correspond to anything real. The statistical contortions only show that in a random collection of questions, some will correlate with any random assignment of people to a category.

That is not to say that having individuals assign items to categories is an unacceptable method. Using individuals who have been trained and whose characteristics as apparatus are measured and known can be a very good system for making assignments. The problem is not with individual judgement, but with uncalibrated judgement. For example, industrial poultry plants need to separate male chicks from female chicks well before they develop any obvious differences. This is done by professional chicken sexers, who process large batches at a time, thousands a day. The chicken sexers can be checked by waiting for their batches to grow a few more weeks, when the difference between males and females becomes obvious. This is a task where untrained humans perform about as well as flipping a coin to decide if a chick is male or female. Trained sexers, by contrast, are around 98% accurate.[6]

Similarly, in psychiatry, the diagnosis of depression can be made in a number of ways. The quickest and cheapest method is to use a questionnaire such as the PHQ-9.[46] However, the primitive notion such questionnaires are trying to measure is not defined in terms of a questionnaire, or even in terms of a structured interview by a trained professional, which is how such questionnaires are calibrated. Instead, both the questionnaire and the structured interview are measuring a primitive notion that is defined in terms of a cluster of clinical outcomes ranging from lack of energy and inability to hold a job to self harm and suicide.

Primitive notions not only need to be operationalized, they need to be operationalized explicitly. If the operationalization is implicit, practitioners are left with only an appeal to authority for which primitive notions are "scientific" and which are not. This statement from Lauren Brent about the study of friendship in primates is a very clear example of this happening:

> Until very recently, and we're talking this year (2012), we didn't like to use the word "friendship" when we were talking about serious science. In passing, speaking with our friends and colleagues we would talk about friendship in other animals, but we would

never write it in a scientific paper.[12]

Like idealized trials and reasonable practice, primitive notions change over time, usually as the idealized trials that define them shift. We saw such a shift in chapter 2 in considering Worthington's switch from beautifully symmetric splashes to photographic plates showing lots of variation.

Another nice example of primitive notions changing with idealized trials is the history of the notion of pressure in physics. In classical thermodynamics, the pressure on a surface is treated as a fixed value. As microscopes made ever smaller experiments possible, the pressure on tiny surfaces was found to vary randomly around the value measured in large experiments. To go with this change in trials, pressure was reconceived as being the force exerted on a surface by randomly moving molecules bouncing off it. Values of the primitive notion changed to being a probability distribution with certain properties rather than a fixed value.

The idealized trial used to define a primitive notion can also remain unchanged, but be interpreted differently, as happened with mass. For most of the history of physics, the mass of an object was regarded as fixed, but, in the early work on special relativity, mass was redefined to increase as the velocity of an object approached the speed of light. This was consistent with all of the previous uses of the term, since none of the idealized trials for measuring mass were practicable for an object moving at such speeds.

However, that was not the only possibly choice. The mass could be declared to be the mass of the particle at rest, and the formulas for energy and momentum modified instead, which is what physicists use today. The field switched because the mass at rest of elementary particles is one of the primary characteristics that experiments measure and theories calculate in particle physics, so it was more convenient to define it as the mass.

The shift in the definition of mass was not the result of work occuring over a year or even a decade, nor was it the result of one person's efforts. Even the shift that resulted from Worthington's work on droplets was the result of several years of work. Practitioners rely heavily on an unconscious familiarity with their primitive notions, and change them slowly if at all. Richard Hamming noted of Einstein on quantum theory, "the person who opened up the field did not understand what he had done, and [was] best ignored at a later date."[34] And Max Planck wrote in his autobiography, "A new scientific truth does not triumph by convincing its opponents and making them see the light, but rather because its opponents eventually die, and a new generation grows up that is familiar with it."[55]

The slow change in primitive notions need not arrive at something usable, however. Sometimes the primitive notion is found not to correspond to anything and is abandoned entirely, or reduced to a set of labels for organizing books. The concept of species underwent this process. Species is a holdover from a early biological conception of species as fixed things established at the creation of the world. Unfortunately, the boundaries between species are at best murky. Pacific Salmon fishermen will tell you that they regularly haul up salmon that are somewhere in between the six accepted species.[57] Species whose range encircles some large obstacle, such as the Himalayas, often form what is called a "ring species:" the species varies gradually around the obstacle until it wraps back on itself as something sufficiently different that it is no longer the same species.

The situation in microbiology is even less clear. There were huge debates through the 19th and 20th century over what constituted a species in bacteria. The dispute was never settled. The participants simply became exhausted. There are still bacterial taxonomists, but they tend to be more worried about the assignment to higher levels of the taxonomy rather than where one species ends and another begins. Other microbiologists have relegated species to a label they write on their tubes and plates for storage, or a label associated with a medical diagnosis (anything that has these chemical characteristics and causes these symptoms in a patient will be classified as *Pseudomonas aeruginosa*, or, if it's a little too far off for the microbiologist to feel comfortable calling it that, as *Pseudomonas sp.* (a Pseudomonas species), which is effectively the same thing. Since it's based on the uncalibrated impression of this or that practitioner we can hardly consider it a reproducible trial. Species has ceased to be a primitive notion, and become a label for whether some collection of trials is reproducible with two different group of organisms.

The concepts of trials, idealized trials, reasonable practice, and primitive notions give us the apparatus to understand the trials we observe practitioners running. A practitioner runs a trial to measure the value of some primitive notion. The trial is designed to approximate some idealized trial, and whether the practitioner's community accepts the measured value is based on a shared standard of reasonable practice.

All of these pieces change over time. Practitioners develop new idealized trials and refine existing ones. The standard of reasonable practice tightens as practitioners discover errors that invalidate trials. And as idealized trials trials or their interpretation shift, so do primitive notions.

But how do practitioners think up these trials?

5 | Generating trials

How do practitioners generate trials to run? If we look at the trials a particular practitioner runs, they are often very similar. The bacterial geneticist is largely generating mutations, selecting or screening for ones with a particular characteristic, and then complementing the mutations to make sure what she found wasn't an artifact. The means of mutagenesis changes, as does the method and target of screening or selecting, but there is an underlying template common to all such experiments.

These templates shows up again and again. Idealized trials are not built from scratch each time. Practitioners have a template in their head that they specialize for the purpose at hand. This works the same way as for any other form of expression. When writing, we depend on templates to provide efficient means of expression. I have found that students who are talking their way around what would, for me, be a simple statement are inevitably missing a rich set of templates to work with. Similarly, the templates a practitioner depends on are one of the most important parts of her training. It makes it possible for her to both generate idealized trials of a form that she knows how to approximate reasonably, and also to understand trials that others in her field have run by recognizing the template that they were working from.

A bacterial geneticist reading papers in her field recognizes the mutate and select template. When she is looking at the description of an experiment, she is looking for what gets plugged into the template's holes: here is the selection, here is the mutagenesis method, here is the complementation. Someone for whom that template is not on tap must slowly try to work out the logic of the trial. If their training is in too distant an area, they may not be able to from the published description.

There are certain aspects or building blocks of templates that show up repeatedly in different fields, such as positive and negative controls. A negative control is a condition added to the trial which should produce a negative result unless something has gone wrong to make our approximation unreasonable. A positive control is one that should produce a positive result under the same conditions. For example, if we are testing a new pesticide, we don't spray some of the plants (a negative control), and some others we spray with a pesticide that we know works (a positive control). If the negative control plants don't get eaten, then we know that our experiment isn't telling us anything about the effectiveness of the pesticide. If the positive control plants do get eaten, then there's something wrong in the other direction.

We see the same building block in the trials of a bacterial geneticist. When creating mutants, some of the bacteria are handled in the same way as the others, but with no mutagen. Any results from them in the selection is due to the background rate of random mutation. That is the negative control. Similarly, a mutant expressly created to be selected by the screen is created and used to make sure the screen is behaving as expected. Physicists tend to call their controls "calibration": turn the detector on but not the experimental source and make sure its empty state behaves as expected, then take a known source, such as a hunk of radioactive cesium, and make sure that signal shows up the way you expect.

Now, a practitioner does not consciously reach for this or that template. The actual generation of a trial is an emotional event. That may seem very odd given the view practitioners have of themselves as rational, but

> [E]motions are actually the internal architects, conductors, or organizers of our minds. They tell us how and what to think, what to say and when to say it, and what to do. We "know" things through our emotional interactions and then apply that knowledge to the cognitive world. Let's look at some examples of how this works. Consider how a young child first learns how to say "Hi!" as he greets other people. A toddler doesn't memorize lists of appropriate people to say hello to. He merely connects the greeting with a warm, friendly feeling in his gut that leads him to reach out to other people's welcoming faces with a verbalized "Hi!" If he looks at them and has a different emotional feeling inside, one of wariness, he's more likely to turn his head or hide behind your legs.[30]

For adults, it is more complicated. We have much more complicated patterns of speech available to us, patterns which are adapted to fill particular needs. There is no explicit, computer-like choice among the patterns. Instead, faced with an emotional impulse to expression, lots of patterns begin to activate, but most of them are suppressed as inappropriate to the impulse. Indeed, all of them may be suppressed, leaving you struggling to articulate yourself. Eventually you settle on one.

Templates are just such patterns, so there is an emotional criterion that drives selection of an idealized trial. For some trials, the emotion may be wanting to satisfy a boss or a mentor, to impress the girl in the next lab, or to show them, show them all.[1] But it need not be. The emotion can be about the generated pattern itself rather than any externalities, and for practitioners

---

[1]Muwahahaha.

who are generating trial after trial it usually is. We have a word for a criterion that drives us to choose this pattern over that pattern for reasons due only to the pattern itself: aesthetics. A practitioner generates a trial to satisfy her sense of aesthetics. Mathematicians call it taste, and regard it as the essential trait that a mathematician must have. For all practitioners, an important part of their training is to internalize the aesthetics of their field.

The first idealized trial so expressed may not be aesthetically satisfying as we wish, just as we may wish to restate something we have said more clearly. So we examine what we have produced, find the places in which it fails for its purpose (and is thus aesthetically displeasing for a practitioner), and then are driven to express a new variation on it.

Dealing with someone who hasn't acquired the same aesthetics is frustrating. I remember talking with biologists after I had been trained as both a biologist and a physicist. When something would offend my aesthetics as acquired from physics, I would protest, and my colleagues would ask why. I was often left sputtering. The only answer that immediately came to mind was that it was ugly, though given some time I could usually articulate an actual problem that I was seeing. Fortunately, I had a masters student working with me who knew a bit of both sides, but was less indoctrinated in each, and who would translate for me while I mumbled incoherently.

The aesthetic is at work in recognizing trials as well. Just as when you are reading a logical argument, your mind processes the logical templates such as "if this, then that" or "this condition is equivalent to that condition," a practitioner seeing a trial of a form encompassed by the aesthetic in which she has been trained has the corresponding pattern activated in her mind, and the corresponding emotional reaction.

The trial that a practitioner expresses doesn't encompass every mathematical detail required to approximate it. What levels should be used for this factor? How many subjects do we need to see this effect size? What is the smallest effect size we can consider important? These are details of the approximation, though the aesthetic will often suggest the method that should be employed to answer these questions.

The number of templates a practitioner internalizes depends on her field. History is constantly looking for new idealized trials, new ways of interrogating the historical record. There is only one historical record. Trials cannot be repeated on independent data. Instead, another trial must be dreamt up that produces the same result by different means. This is true of most of the social sciences. At the other extreme, physicists, chemists, and laboratory biologists have relatively few templates. Molecular biologists use basically one template for idealized trials. Practitioners in these fields can repeat the same trial, and so they spend little time worrying about the form of their idealized trials.

This parsimony leaves these fields somewhat at a loss to assimilate work that doesn't fit their handful of templates. For example, take the evolution of copackaging studied by Sachs and Bull.[60] They took two species of bacteriophage,[2] neither of which kill their host cell. One of the species of bacteriophage had been genetically modified to render the cell it infects immune to the antibiotic kanamycin. The other had been modified to render its host immune to a different antibiotic, chloramphenicol. The two species were mixed and added to cultures of uninfected cells, and, after allowing time for them to infect the hosts, both antibiotics were added to the culture, killing all the cells that were not infected by both species. Then the culture medium of the cells was replaced to remove the virus used in the initial infection, and, after waiting some time to let the infected cells

---

[2]Viruses that infect bacteria

generate new virus, the new viruses were harvested and used to repeat the process on a new set of uninfected host cells. The goal of the experiment was to see how two species evolve if they are forced to propogate together.

They found that over successive generations the genomes of both viruses ended up packing in the protein capsid of one of them, and, after this copackaging, the genome of the virus whose capsid was abandoned began to steadily lose genetic material, until only three coding regions remained by the end of the study.

Most molecular biologists, faced with this, agree that it is very interesting, but have no framework to make use of it. The material is that of evolutionary theory even though the methods are molecular biology. Would they copackage again if the experiment were repeated? Would the two viruses copackage into the other virus's capsid next time? Would it take the same number of generations to copackage? Would the virus that stops producing its own capsid lose genetic material in the same order? We have no idea. These are questions that evolutionary biologists typically don't bother asking since they do not expect to be able to repeat a piece of evolution. But the methods are molecular biology, and we could repeat it, and, though these authors did not, other authors have, particularly practitioners studying industrial cell culture in chemostats. For example, a particular strain of *E. coli* in a chemostat will undergo a fairly fixed set of mutations as it adapts to the conditions, and the rate of those mutations appearing turns out to be independent of the rate of growth of the bacteria in the chemostat, for reasons that are fascinating and subtle.[72]

On the other hand, there are trials of this kind that are so long that they are unlikely ever to be repeated, and so must be dealt with as natural history. In particular, Richard Lenski's group has been passaging *E. coli* in chemostats for decades, freezing samples along the way.[47] Then they go back and study the samples as though they were a fossil record. Lenski's work is as puzzling to molecular biologists as that of Sachs and Bull.

So, both generating and understanding trials are based on templates in the practitioner's mind. The triggers that select among templates are emotional. A practitioner generating a trial selects among templates based upon an emotional reaction to their fit to the problem at hand, that is, an aesthetic reaction. A practitioner reading about a trial tries to fit it into templates she understands. Internalizing templates and a corresponding aesthetic is one of the most frustrating parts of learning a discipline. It often takes a student hours of staring at the same thing to understand it, while an accomplished practitioner glances over the description and understands what she sees in a manner of minutes. But it is that frustrating work that internalizes the templates and makes the practitioner's fluency possible.

Similarly, because they are based on unconscious mechanisms acquired through hard, frustrating work, the change over time in templates is one of the hardest modes of progress for a practitioner to envision. A practitioner designing an agricultural experiment today must consciously pause and work to generate a trial not involving randomization. A practitioner in 1900 would never do so. Old practitioners who have not adjusted steadily over the years often find themselves even more frustrated when trying to understand new work in their field than a student would be since the templates they internalized many years ago may have shifted out of recognition. This difference in templates and even the difference in the number of templates practitioners in different fields employ is also the first hint we have of how to explain why a practitioner finds fields other than her own alien.

6 Through the looking glass

So far we have made no mention of anything resembling a theory. We have only discussed particular trials, how they are generated, their form, and how they are accepted as reasonably conducted or not. We will now rectify that omission, but those looking for deep philosophical claims for theories are going to be very disappointed.

A theory is a formal set of relations among some set of quantities. The theory of classical mechanics is a set of relationships among the quantities mass, position, momentum, acceleration, force, and time. History is a set of relationships among events or conditions in the past, whether temporal, causal, or otherwise. A theory is often not couched in mathematics—theories in history, anthropology, archaeology, and biology, to name a few, rarely are—but it is usually supposed to be rigorous enough that it could be couched in mathematics. In fields with complicated relations of only a few kinds, such as physics, theories tend to be treated mathematically. In fields with fairly simple relations, but many of them of many different kinds, such as history, the cost of erecting symbolic formalism for so many distinct cases isn't justified when they can be handled perfectly well without it.

What all theories have in common, whether couched in mathematics or otherwise, is a set of terms not defined in the theory. Classical mechanics makes no mention of what mass or momentum are. Pharmacology does not define death when measuring what dose of a chemical is lethal. These undefined terms are given meaning by primitive notions from trials. This usage is also what led me to call them "primitive," since they form the irreducible components of a theory.

A theory, then, is a formal set of relationships among primitive notions that is meant to recapitulate the values taken on by those primitive notions in trials. It is as though we were constructing an image in the looking glass of the reality of the trials.

As an aside, it is important not to add phantoms in the looking glass by imputing to the primitive notions in a theory more than is actually operationalized. In quantum mechanics, van Kampen put this as "Whoever endows [the wave function] with more meaning than is needed for computing observable phenomena is responsible for the consequences."[41]. That isn't to say that there is no place for further interpretation of theories. For example, Bell credits the pilot wave approach to quantum mechanics as the tool that made it possible to figure out his famed locality inequalities.[3] But it is important to understand what is interpretation and what is essential to the theory. Difficulties with interpretation may be brushed aside. Difficulties with the theory may not.

A theory, as a set of formal relations, a mirror of trials, is never "correct." There is no mystical ur-theory under reality that we are trying to uncover. A theory is a human construction. It may not be correct in some Platonic sense, but it can be accurate to some degree. That degree may be extremely high. The last time I checked, the gyromagnetic ratio of the electron predicted by quantum field theory was accurate to twenty one decimal places, as far as has been measured at this point in time.

But even saying that a theory is accurate is not always straightforward. Consider an historical account. If I say the Americas were populated by people who crossed from Siberia during an ice age, then spread south, and formed a variety of societies, that is accurate, but most would find it unsatisfying. On the opposite end of the spectrum, most people would find a forty volume history of the Americas equally unsatisfying simply because they have neither a framework to handle that level of detail nor the time to read it. Such a huge amount of data would need to be further analyzed and reduced to be useful for most purposes, which is how practitioners use such detailed accounts as the United States census[68] or the Sloan digital sky survey.[1]

There isn't a single, acceptable level of accuracy even for a particular practitioner. The same physicist can happily use a detailed molecular model of water and a crude model of magnetism in solids for different purposes. The important part is that a theory be accurate enough for the purpose at hand. This means that a theory must come with some assessment of its accuracy, usually in the form of a summary of the trials it recapitulates and those that it doesn't. The theory of beam deflection used in mechanical engineering comes with detailed measurements of what range of stresses and deflections the theory works over for a given material. A microbiologist keeps a set of strange organisms such as Walsby's square archaeon[9] in his memory and tests theories against them.

Theories, like trials, are generated according to templates in a practitioner's mind. For example, in 1817, the chemist Döbereiner noted that strontium, calcium, and barium have similar chemical properties, and the atomic weight of strontium is the average of the weights of the other two. In 1829, he found several other such triads, and the field of chemistry was dominated by increasingly complicated attempts to organize the known elements in terms of triads dominated until the 1860's. Then larger periodic systems such as those of Mendeleev emerged and replaced triads as a theoretical gambit.[61]

As another example of a template dominating a field for a time, in 1913 Niels Bohr introduced a theory that recapitulated the spectrum of light emitted by hydrogen. His theory consisted of taking Rutherford's model of the atom as electrons orbiting a nucleus, and saying that the electrons orbited only in one of a fixed series of possible orbits that followed a simple mathematical pattern chose to match the hydrogen spectrum. Though utterly *ad hoc*, this gambit fit the data for hydrogen beautifully. Bohr also used it to account for the spectrum of ionized helium, which also has a single electron, and Sommerfeld introduced relativistic corrections to get an even better fit for hydrogen. Unfortunately, it failed miserably for every other problem it was applied to, but for a decade after Bohr introduced his model all attempts by the physicists of the community were variants of his model. Finally Heisenberg, Jordan, and Born introduced matrix mechanics in 1925.

What can we say about the templates that practitioners depend on to generate theories? First they are often explicitly constrained. For example, physicists insist that their theories not depend on the system of units they are expressed in. For example, if a theory uses the equation $a = b + c$, the quantities that variables a, b, and c correspond to must have the same dimensions. For example, they could all have dimensions of length or of mass, but a physicist would immediately reject any theory in which a had dimensions of mass and b and c had dimensions of length. For example, if $a = 3$ pounds, $b = 2$ inches and $c = 1$ inch, the equation is true in those units, but if we convert the quantities to $a = 1.4$kg, $b = 5$cm and $c = 2.5$cm, the equation is no longer true. Reasonable practice in the field demands that a theory be true independent of the units used to express its primitive notions.

In the same vein, practitioners of biology will not accept teleology in evolution, that is, something cannot be said to have evolved a feature because the final stage of the evolution is beneficial. Evolutionary explanations must be local, phrased only in terms of populations that exist at each point during the process. We cannot say that some early creature evolved light sensitive spots so that its descendants would have eyes. Any explanation for the appearance of the spots must be solely in terms of the creature, not its descendants. Similarly, in history we can't claim any effect of the future on the past. An elementary form of this constraint is a practitioner rejecting the statement "Charlemagne had to win his battles, or there would be no France." The existence of France is not inevitable.

Beyond such field specific constraints, theories take a more limited set of forms than trials since they are, in the end, always a set of relationships among symbols, as opposed to all the richness of dealing with the real world. This means that we can enumerate their modes of progress more precisely that we can for trials.

The first, obvious mode of progress is extending the reach of a theory, making alterations to it to let it account more trials than it did before. Perhaps the simplest example is writing histories of areas that had not previously been dealt with. There are a surprising number of thesesuch blank spots on the map. The history of Africa for the past few hundred years was largely written about Europeans in Africa. The past few decades have seen the emergence of a history of Africa itself, including fascinating work on human migrations and cultural interactions inferred solely from the linguistic record.[18]

Progress in the explanatory power of a theory is often driven by a large amount of new data. The great excitement and activity in particle physics in the 1960's and 1970's was largely the result of a huge amount of new data that had to be explained, and an enormous amount of work went into formulating theories to account for the trials, culminating in the Standard Model of Particle Physics.

Our second mode of progress is unifying two theories into one.

Given two theories that apply in different domains, it is often fruitful to combine them and see what emerges when the factors that each account for occur together. The resulting theory is usually more complicated than those it unifies, but sometimes gains a richness that justifies the complication.

For example consider the unification of the theory of evolution and the theory of games in the 1970's[29]. Game theory provided mathematical tools for predicting the behavior of actors in a situation if you knew their available actions and what they valued. Combining them, with actors playing multiple rounds of a game and being evolutionarily selected between rounds, produced a thory of processes that generated sophisticated, rational behavior even in the absence of intelligence, and provided a language for discussing the selection of behavioral traits such as social structures in a precise way.

Sadly, unification does not always work. There have been numerous attempts over the years to unify geography and history. The simplest position is that environment determines fate. However, this does not account for most of the variation seen in history. Steadily other factors have been introduced and the theoretical apparatus refined to try to make sense of the events of the past.[39] That does not stop people from reviving the idea, whether from racism or incompetence, as Jared Diamond did in his book *Guns, Germs, and Steel*,[15] but there is no simple unification that subsumes both.

Our third mode of progress for theories is rather different from the extension and unification. The others modes of progress are about increasing the reach of science. This one is about increasing the reach of individual practitioners: the simplification of theories. Any individual has limited reach. As my friend Jeff Gaynor put it, "You are a mid-sized primate who can, in the right circumstances, pull off the minor miracle we call Science. Kewl pet trick, ain't it?" As a field accumulates material, individual practitioners can either specialize or the material can be condensed to fit into one person's head. Specialization has its own perils, since solutions to common problems must be reinvented by each specialization instead of being shared by the whole field, which slows down science as a whole. So condensing existing material for faster learning and use is an important activity if science is not to grind to a halt from the sheer inability of a practitioner to know enough to do anything that hasn't already been done elsewhere.

There are two ways to simplify. The first is to specialize the material. Thus physicists are taught courses in mathematical methods, omitting much of the rigor mathematicians are concerned with and emphasizing explicit computation. Biologists are taught chemistry with most aspects besides organic chemistry removed. Many practitioners are taught a statistical methods course featuring a handful of techniques for the most common special cases in their field. This may seem a trivial form of simplification, but the ubiquity of books with titles like *Statistics for Biologists* or *Methods of Mathematical Physics* shows how highly valued it is by practitioners.

The other way to simplify a theory is to actually reduce the difficulty and complexity of the material without changing its reach. This is a much respected activity in mathematics, where a simpler proof of a known result is considered important progress. Nor is mathematical simplification the only way to reduce complexity. Einstein's special theory of relativity was not mathematically novel. Lorentz contraction, a constant speed of light, and all the other machinery of it had been put together by various people over the preceding decades. However, all those pieces were wrapped up in an increasingly complicated set of ideas around a hypothetical medium called "ether" in which light propogated. The theory of ether worked to predict and reproduce the results of trials. The special theory of relativity's novelty was in postulating a slightly different set of kinematics which reproduced the same results, often via the same mathematics, but with a vastly simpler structure.

Such simplification can go too far. In the late nineteenth century, there was a general belief that bacteria changed form, and what we now recognize as very different species were variations of the same organism. Now, most of the observations of "changing" form were due to contamination with other species with different shapes, since the techniques for maintaing a pure culture had not been invented yet. Cohn and Koch developed many of the modern tools of culturing bacteria without contamination, which led them to the opposite extreme: every bacterial species has a fixed form. Today we regard bacterial form as somewhere in between. Most species have only a few basic forms that correspond to different parts of their lifecycle, though there may be variation in the details of that form. *Escherichia coli*, for example, has a single form of a round ended rod with flagella trailing off of it, but its width and length depend on the richness of the medium in which it divided from its parent. *Bacillus subtilis*, on the other hand, is a rod in rich conditions, but in hostile conditions produces small, round spores. Mycobacteria are generally squiggly sausages, but when damaged by antibiotics will start to produce branches and tree shapes. This is hardly the morphological flights of fancy of the mid-19th century, but it is more complicated than Cohn and Koch's model.

Another aspect of simplification is being able to recall material more easily, or, as Richard Feynman put it, "the real glory of science is that we can find a way of thinking such that the law is evident."[22] The best view of this I know of is from *Mathematics Under the Microscope*, wherein the author contrasts the idea of a proof with that of a recovery procedure. A proof is an argument couched in some form of logic. A recovery procedure is a view of something that has a few, memorable bits that constrain everything else into a unique form:

> It is time to fix some terminology. A *recovery procedure* is a set of heuristic rules which we vaguely remember to apply when we want to recover a mathematical fact. It is like a poster on the control panel of some serious machine, a submarine or a plane, which says what the crew should do when things have gone haywire. A *rederivation* is a semiheuristic argument made in accordance with the recovery procedure.[11]

Finally, the constraints on theories develop over time. Prior to the development of thermodynamics, arguments for or against the existance of perpetual motion machines were taken seriously.[64] The development of that science settled the question once and for all: they cannot exist, and the reasoning why is embodied in the second law of thermodynamics, which is probably the strongest constraint in all of physics. As Eddington put it, "if your theory is found to be against the second law of thermodynamics I can give you no hope; there is nothing for it but to collapse in deepest humiliation."[17] Similarly, any biological theory that depends on the spontaneous generation of organisms from inorganic matter to explain present day phenomena is immediately disregarded, despite the prevalence of such ideas in biology prior to the 20th century.

Constraints need not always tighten. In some cases they relax. For example, until the 1950's physicists assumed that rerunning a trial with right and left swapped in its construction would behave equivalently to watching the original trial in a mirror, what is known as parity. Then in 1956, two experiments, done at the suggestion of the theorists Lee and Yang, found that parity was violated. Chien-Shiung Wu found asymmetry in the emission of electrons from unstable cobalt nuclei, and Leon Lederman found an asymmetry in the decay of pions and muons.[51] These results were a severe shock to the physics community, and began a hunt to outline under exactly what conditions such seemingly fundamental symmetries are violated, a hunt which continues to this day, relaxing charge symmetry and time reversal symmetry in the process.

To summarize, a theory is a formal structure which relates the values of some set of primitive notions. Its usefulness is judged by how well it matches the relationships among the values measured in trials, and this formal structure provides a way to organize and summarize large numbers of trials. Practitioners make progress on theories by extending the range of trials that the theories summarize, by unifying two theories, by simplifying them to reduce the effort required of other practitioners to use them, by elucidating errors to tighten the constraints on a theory, or by finding holes in existing constraints that force them to be relaxed.

7 Disciplines

At this point we have all the pieces of our model in place. Practitioners generate trials to measure the values of primitive notions, and try to build theories that recapitulate the relationships among the values measured in trials. The trials are generated to approximate an idealized trial, and practitioners use a shared standard of reasonable practice to judge if the approximation is adequate.

With this machinery in place, let us look at the aggregate of trials, primitive notions, and theories that a practitioner uses. The first thing to notice is that the primitive notions, idealized trials, and theories that practitioners in different fields use barely overlap. A particle physicist's various trials measure the values of a small number of primitive notions, largely the number and individual masses of the particles that result from an interaction among particles, the direction and speed of the particles resulting from the interaction, and the relative frequency of different decay products that the interaction can produce. An historian has none of these primitive notions. A molecular biologist shares no primitive notions with an economist. An ecologist and a chemist may share a few primitive notions of measuring concentrations of a chemical, but their theories and trials are utterly different.

These distinctions form gulfs between practitioners of different fields. An historian has internalized her field's reasonable practice, primitive notions, and templates for idealized trials and theories. When faced with the corresponding material in astrophysics, she may as well be a layman. The upshot of this is that disciplines are real things, not just academic departments.

If disciplines are real, then we would like to characterize them, to understand the form they take. Characterizing disciplines suffers from the problem of all observational social science of what to include and what to exclude. The most useful characterization that I have found is to look at what a practitioner of a discipline means when she says that she understands something; that something is interesting; and that something is beautiful.

When a practitioner says that she understands a phenomenon, she means that she has trials of that phenomenon that produce values of her primitive notions, and that she has a theory that reproduces the results of those trials. Thus looking at what a practitioner says that she understands and does not understand will directly show us the idealized trials, the primitive notions, and the templates of theories in her field. Consider what practitioners of different disciplines mean when they say that they understand the difference between the smooth and wrinkled strains of peas that Mendel used in his original genetics experiments.

When a geneticist says that she understands this phenomenon, she means that she knows that, if she crosses two plants with a known ancestry, she can predict the probability of getting certain numbers of wrinkled or smooth offspring in succeeding generations. We read off from this that she has an idealized trial of the form of a breeding experiment; she has a primitive notion of probabilities of traits manifesting in offspring; and she has theories that relate those probabilities across generations of breeding.

A developmental biologist would say that she understands the peas when she can describe how the peas grow from a few cells into macroscopic objects with or without wrinkles, and can intervene at various stages to force them one way or another. An anthropologist or historian would say that she understands the peas when she can produce an account of when the strains arose, where, and in what contexts. A chemist would say that she understands the peas if she knows their difference in composition, whether one has more sugars, more water, or other differences at the chemical level than the other. Peas aren't an object expressible in a physicist's primitive notions. The water absorption of smooth versus wrinkled peas is expressible, or how tightly the two kinds will pack together. Similarly, an economist might be interested in the return on strains of different kinds of peas. Many of these meanings of understand overlap only insofar as they are about peas.

Once we know what a practitioner means by understand, next we ask what she considers interesting. Interest is determined by how much effect something has had or that we expect something to have on the theories in a discipline.

For example, for most of the history of biology, trees of descent showing the relationships among species were built based on phenotypic characteristics such as skull shape or the presence or absence of feathers. There were parts of the tree where practitioners weren't sure what the relationships should be. Since traits that a biologist can easily observe can also be easily selected for, we don't know if two organisms have the same trait because they were both in environments that selected for it, or if they both have it because they are closely related. If you measure many traits then you should get similarity from selection instead of descent in only a few of them. But how do you build trees including both animals and bacteria? The traits that you would measure on an animal do not exist for bacteria and vice versa. DNA sequencing provided solutions to both. There are loci of DNA that encode molecular machinery shared by all cellular organisms, and each base pair of the DNA is a separate trait.

Carl Woese was the first to construct a full tree of life based on DNA sequencing[73]. It was interesting beforehand because biologists expected it to resolve many of the ambiguities in the previously constructed trees. What he found was actually even more interesting. Biologists had expected the tree to have two branches from its root, one for bacteria and one for eukaryotes. Instead, it had three. The bacteria were split into bacteria and archaea.

Similarly, until the 1960's, neuroscientists believed that once a mammal reached adulthood, it did not grow additional neurons. But in 1967, Altman and Das[2] discovered that brains continue to develop new neurons throughout life. This changed the constraints on theories of learning in neuroscience almost immediately. Such a large change made it very interesting.

Before a discovery, "interesting" involves not only how much effect it would have on the theories of the field, but how likely practitioners think it is. Before Altman and Das, neuroscientists thought that neurons did not grow during adulthood, but it was possible that they were wrong so experiments to test that were interesting. On the other hand, anyone trying to build a perpetual motion machine to violate the second law of thermodynamics is greeted with a yawn. It would certainly be a revolution in physics, but no one expects it to happen.

Before we leave what it means for something to be interesting, a caution: when studying what is interesting for a discipline, we must distinguish between practitioners of the discipline and those who use the discipline as a tool. A microbiologist finding that a strain of bacteria isolated from a cow's gut is resistant to penicillin may be quite important for agricultural policy, but it is not particularly interesting biologically. Livestock are often fed prophylactic antibiotics, their guts are full of bacteria, and bacterial populations exposed to antibiotics evolve resistance. Similarly, measurements of how far different kinds of wood deform under a given stress is of little interest to a physicist, but of great interest to a structural engineer. Spintronics, the construction of circuits that work with the intrinsic magnetic moments of the atoms involved instead of sending electrons along wires, doesn't have any particular implications for physics, but is very interesting for computer engineering.

Finally, what do practitioners of a discipline find beautiful? The old saw that "beauty is in the eye of the beholder" is quite true, and we can use it to find out something about the beholder, in this case a reasonable practitioner of a discipline. To figure out what it shows us, consider how a practitioner's develops her aesthetic.

Aesthetics in the sciences are largely a matter of training. Science is not an instinctive activity, so we need not consider elements outside of training. During that training, practitioners are shown examples of earlier work. What a practitioner passes on to students is a combination of what she learned as a student that has not been overturned in the intervening years, along with a smattering of new material that she feels is important. Thus the majority of what a student receives is material that has held up over time. What is beautiful becomes a proxy for how robust a piece of work is likely to be.

This is particularly clear in microbiology. There is a special place in a microbiologist's heart for work carried out with nothing but organisms grown on Petri dishes, transferred from Petri dish to Petri dish with toothpicks, examined with low power microscopes, and, if the organism causes a disease, inoculated into an animal. These tools were developed in the 1880's by Robert Koch, and his work using them to isolate *Mycobacterium tuberculosis* and show that it caused tuberculosis[44] remains as reasonable today as it did when he performed it. Generations of microbiologists have learned this and then passed it on to their students.

This aesthetic preference is reinforced by new work using the same tools. In 1988, Weinert and Hartwell used the same tools to demonstrate the existence of mechanisms in eukaryotes that arrest cell division in the presence of DNA damage, but are not part of the chromosome replication machinery, what are referred to as "checkpoints."[70] This work has joined Koch's as something taught to students.

Checkpoints also provide an example of how specific to a discipline an aesthetic is. Biologists have assembled a detailed list of checkpoints in the cell cycle, and mentally think of the cell cycle as cleanly proceeding or stopping like cars at a stoplight. This same view makes a physicist recoil. Physicists have strong aesthetics around the form of their theories. The models of molecular processes that have survived to be included in their training are all in terms of noise and randomness, and physicists who have transitioned to working on the eukaryotic cell cycle have started to replace the simple checkpoint model with more detailed, physically accurate models based on coupled molecular oscillations.[49]

The meaning attached by a practitioner to these three words—understand, interesting, and beautiful—yields a strong sketch of a discipline. "Understand" tells us about the trials, primitive notions, and theories of the discipline. "Interesting" tells us what work has had an important effect on the field's progress, and what work practitioners think is likely to have an important effect. "Beautiful" tells us about the work that has endured and forms the foundation of how practitioners view their field.

Given this sketch, we can look at how the forms of different disciplines are shaped by their topics of study, and why they are so different.

Physics studies a fairly limited number of primitive notions such as position, momentum, temperature, charge, mass, and the arrangement of atoms in a material. A physicist's trials are all controlled, repeatable experiments. Anything which depends on historical accidents and thus does not fit this mold is shoved into a subfield such as astrophysics or geophysics. Since the trials can be repeated and the number of primitive notions is small, practitioners have been able to develop the theories relating them to a high degree of complexity.

Contrast this with history, which has no possibility of running a repeatable experiment. There is one historical record, with all its accidents and systematic errors built into it, and, since an historian can never independently repeat a trial, she is always trying to craft new primitive notions and new idealized trials to try to confirm earlier work via some other part of the historical record. A field with an unknown and continuously increasing number of primitive notions and idealized trials, and where trials cannot be repeated, won't support a complicated theory relating particular primitive notions. There is no way to know whether the additional complexity put in to describe the one instance of the trial that is possible describes something general or is a result of some confounding, historical accident. Simple models work better. Nor, given the increasing number of primitive notions, does it make sense to spend time building a complicated model for a particular, small set of them.

Physics and history hardly resemble one another. Nor, given the character of their different subject matters, would we expect them to. We expect to find the same idiosyncracies in any other discipline. Epidemiology could, in theory, be an experimental science, but everyone regards releasing real plagues to watch them propogate as unethical, so it is a purely observational science. Economics, like history, works on observations of a wide variety of primitive notions, but those notions all measure numbers. Thus its models are mathematical, even though, like history, its observations will only support simple relations among primitive notions. This leads to the apparently ludicrous situation of a mathematical field where mathematical sophistication leads to nonsense.

Biology is a particularly odd case. It is actually three disciplines inextricably intertwined. The first discipline is the study of inheritance and transmission of traits, which we call genetics. Its practitioners work in terms of breeding experiments, that is, if I begin with these strains and cross them, what is the assortment of traits in the offspring? The second discipline is the study of the organism as physical object, more or less physiology. It includes anatomy and biochemistry, biophysics and biomechanics, all the study of the physical being of organisms. The last discipline is the study of the lifecycle of organisms, of their relationship to their environments, and of their descent from earlier organisms. I usually refer to it as natural history, though typically what is called natural history today also includes a certain amount of geology and environmental chemistry.

Any piece of work in biology usually focuses on one of these three, but it must meet the standards of reasonableness in the other two as well. An example of work going awry because it did not deal with all three aspects of biology arose one evening in a talk I attended at the New York Academy of Sciences. A young practitioner, trained as a physicist, and newly switched to biology, had done a set of experiments on the activity of transcription of different loci of DNA in *E. coli* under two conditions: warm with low oxygen to imitate the gut, and cool with high oxygen to imitate the outside world. His results weren't particularly compelling, but during the questions at the end, I asked him what strain of *E. coli* he had used. It turned out that his work was done in MG1655, the most common laboratory strain of *E. coli*. That strain hasn't grown anywhere but a test tube for decades, and is well known in the field to be hyper responsive to anything done to it. There was a sigh from all the biologists in the room when he answered. All the data he had gathered at such expense wasn't worth looking at. He hadn't met the standard of reasonableness in natural history, and all his careful work to meet a reasonable standard in physiology was for nought.

The result of this tripartite nature is that biology is a sprawling field. A particle physicist and a condensed matter physicist can usually understand why the other's trials are run in some particular way. A field ecologist studying food webs and a molecular biologist studying where different mRNAs are localized in a cell can barely talk to each other, and have as much problem understanding why the other's work is interesting as an economist and an historian would.

So disciplines are real things. They differ not only in their areas of study, but the structure that the area of study imposes on the science. Often that structure can be bizarre, as in the case of biology, but it can be described usefully by analyzing what practitioners of the discipline mean by understand, interesting, and beautiful. Working from the material associated with those three words provides a reproducible criterion for selecting what should be part of the description of the discipline.

8 | Student to practitioner

The skills that make a successful student of science are nearly irrelevant to a practitioner. A student is judged by how skillfully she overcomes obstacle after obstacle in a sequence designed by some authority, hopefully accruing some skill in the process. For many purposes this is the right goal. A structural engineer must be able to manipulate classical mechanics with great fluency, but is not expected to reinvent it. An ecologist needs to use the tools of chemistry, but probably isn't much concerned with how to extend them. However, that same ecologist, in her own field, isn't faced with a series of well defined obstacles but an unknown spreading before her with no guide through it.

Transitioning from student to practitioner requires that we know how to approach this unknown. The structure of science described so far is too impersonal to be a guide. "Unifying two theories" and "running a trial" are certainly more concrete than "doing research," but not concrete enough to understand the day to day life of a practitioner. So now we have to invert our point of view from looking back on developments that have already happened, as we have done to put together our model, to looking blindly forward. How do these modes of progress that we see in looking back on the development of a field emerge from a practitioner's activity? How does she organize her activity so that she makes progress?

Let us start by looking at the day to day activity of a practitioner. She comes into her laboratory, office, or field site every day and does fairly menial work, work that largely seems trivial. This triviality is actually important. Unlike a student who is fully engaged in one aspect at a time of the structure of her discipline, so she can develop skill in that one part, a practitioner must question whether something has gone wrong at any level from trial to theory. If what she is doing doesn't seem simple, she is unlikely to be able to handle all those levels. So as a rule of thumb, if you feel impressed at yourself for the power of the mathematics or other techniques that you're using, you're probably doing something wrong.

The meniality of day to day science separates those who "like science," usually meaning they did well in classes in science and are part of social circles where having some knowledge of the results of various disciplines is prestigious, and those who like doing science. Do you enjoy going in and doing the menial work day in and day out? There are many people who loved their course work in the sciences who abhor the actual activity of research. There are others who were marginal in their coursework, but who are superb at this daily grind.

Looking back on this menial activity, we interpret it as "running a trial" or "unifying two theories." The practitioner, looking forward, plans them as such, too. But how does she decide to run a particular trial? When does she make an attempt to unify two theories? To understand this, we need to look at the context a practitioner holds in her mind when she makes these decisions.[1]

The most immediate context is the management of the menial labor. The maize planted in field three needs to be watered again today. That bacterial culture has had its three days of growth in the new medium. This technique for detecting encrypted

---

[1] I am unaware of any literature on this kind of short term planning and execution in the sciences, so what follows is necessarily from my own introspection and from talking with colleagues, with all the problems that self reporting entails.

segments in signals has failed in the way that you thought it might, so try the next approach lined up. Transforming this hunk of DNA into those bacteria failed. Get a fresh aliquot of transformation medium out and let's try it again. These all require focused attention, and in many cases a lot of repetition and tinkering to make something work. We call this material the management context.

But that context does not decide that maize needs to be planted and what schedule it needs to be watered on. The management context sits atop a deeper context of planning. The plans we are talking about are very short term, usually only a few steps into the unknown, and those steps are chosen based on what is interesting. Here "interesting" has exactly the same meaning as when we used it to characterize disciplines, that is, what steps will have the largest expected change in our understanding. "Expected" here is a technical term from the theory of probability. It is best thought of in terms of betting. If a practitioner must place bets on each of the possible steps she can take, what bet should she make? A step may imply a huge change in our understanding if it succeeds, but seem, given what the practitioner knows, to be very unlikely to succeed, and so be a terrible bet. As a ludicrously extreme example, producing antigravity by boiling water in a strangely shaped vessel would be a coup, but no one who knows anything about physics would expect it to succeed. Its poor expected chance of yielding anything makes it an awful bet, and uninteresting as a next step. At the other end of the spectrum, the first time a practitioner makes a step work it may be interesting. Confirming an interesting step a second time is still interesting, since she wants to know that it wasn't a fluke. But every subsequent time she repeats it, it becomes less interesting. It is virtually certain to yield the expected result, but that result has little effect. So reproducing something she has reproduced twenty times before is probably no longer interesting. On the other hand, reproducing someone else's work that she wants to build on is often a very interesting step. The step has succeeded for someone else, which makes its chance of working higher, and it opens the path to other interesting steps.

So what goes into the planning context that a practitioner uses to make these short term plans? The contents are almost a catalog of what differentiates one practitioner in a field from another: the materials she has to hand, the techniques she has mastered, the problems that interest her, the results she is aware of, and the steps of previous plans that have led her to where she is.

For example, consider the following, simplified context. Assume the basic idealized trial of antibiotic research: grow a population of bacteria in some culture, add an antibiotic to the culture, and measure the number of surviving bacteria over time after the antibiotic is added. The population drops precipitously for the first few hours, but then stabilizes at a low level, a phenomenon known as persistence. Then add a second idealized trial: grow a population and add antibiotic as in the previous trial, but, after the population stabilizes, inactivate or wash out the antibiotic, let the bacteria grow up again, and add antibiotics a second time. Measure how the survival is different in the two periods of antibiotic exposure. Typically with persistent cells, identical doses of antibiotic produce identical rates of survival both times.

Let us add a background knowledge of biochemistry. The number of molecules of antibiotic afterwards are the same as the number at the beginning. The antibiotics bind to various molecules of cells. Practitioners in this field also have a whole slew of results in their context about what cellular processes are disrupted by various antibiotics.

Even with the two idealized trials, this practitioner has a huge range of gambits. She can choose the organism to expose to antibiotics, how dense a population to use, what medium to grow it in, the antibiotic and concentration to use for each round of exposure, and how long to let each stage proceed. What particular trials seem most interesting to her depends on the rest of her context. If she has a bent for physical chemistry, she is likely to try varying the levels of antibiotic to see how it changes the fraction of cells that persist, hoping to make theoretical links back to chemistry. If she has a background in medicine, she is likely to measure how combinations of antibiotics change the persistence rate, looking for combinations that reinforce or interfere with each other. A background in natural history would lead her to measure persistence across different organisms.

Joseph Bigger, who first observed persistence and developed the idealized trial described above, had in his context several more facts. First, most bacteria, grown to high densities in a liquid culture, reach a population plateau, called "stationary phase," and in that plateau the antibiotics do not cause a precipitous drop in living population. Second, when cultures grown to stationary phase are diluted into a new culture, there is a delay before they resume fast growth. When antibiotics are inactivated or washed out of a culture, there is also a delay before growth resumes. He proposed a unification that the persistent cells that survived were a subpopulation that were physiologically identical to the cells in stationary phase, which suggested a number of trials (which, in the end, showed that persistent cells and cells in stationary phase are *not* in similar physiological states).

Though a research plan comprehends only a few steps, it may still require an enormous amount of time. The extreme case of this is astroparticle physics, where a single trial may take decades and hundreds of people to build and run. Plans in situations like this are made by committees of senior practitioners, and most of the field works within the plan. Most groups working in areas like this try to be involved in several trials, each at a different stage, so their graduate students can experience the different phases.

Even when the trial is run by an individual, plans may extend over long periods of time. My colleagues who were studying tuberculosis by infecting mice with the bacterium had trials which lasted multiple months. Similarly, agricultural trials require seasons or years to complete. Epidemiologists running longitudinal studies of the long term outcomes of groups of patients have to wait decades for their results. When executing plans takes so long, researchers respond by arranging their work in parallel by working on multiple questions at once. Such fields can be problematic for students trying to transition to being practitioners, since the new practitioner finds herself tackling obstacles from a plan created by someone else, much as she did as a student.

The planning context shapes the management context, and in turn it is shaped by a third layer, a foundational context. This is where we find a practitioner's primitive notions, and theoretical constraints, all the things that shape the gambits in the planning context.

So we have a three layered context that a practitioner uses in her work. She spends most of her time at the level of the management context, carrying out her work. And regularly she hits obstacles. Students often go awry by failing to back off in these situations. They have been trained that they must tackle the obstacle before them. Practitioners have no such obligation. If an obstacle can be cirumvented, all the better. For example, a student told to develop an algorithm to find cells in a given image has a well defined task. If it is difficult, so much the worse. A practitioner can fiddle with how the images are generated and acquired. She can add dyes that don't enter cells, and so stain the background bright red. She can increase the resolution of her images. She can dilute the cells more before putting them under the microscope to make hard to handle regions in the images rarer. These may make finding the cells an easy task rather than a nearly impossible one. And even if these fail, she can fall back to outlining the cells in the images by hand, perhaps with computer assistance to speed up the process. None of these solve the problem of finding cells in the original image, but they all satisfy her needs. On the other hand, tackling the obstacle may be the right thing to do, but an effective practitioner is choosy about which ones she spends hard work on.

Similarly, a practitioner may exhaust the gambits open to her in her planning context, and have no obvious direction to go. At this point she has to "get an idea" or "be creative." That is, she must dredge up something from her foundational context to work on which will open other gambits to her. Now, the foundational context is largely subconscious, so working with it usually requires some kind of trick to unearth material into the practitioner's consciousness where it can be worked on. Like telling a young practitioner to "do research," saying that she must "be creative" is too vague to be helpful. Gian-Carlo Rota in *Indiscrete Thoughts*[58] puts it

> What makes a mathematician creative? Rule one: don't ask him to be creative. There is nothing deadlier for a mathematician than to be placed in a beautiful office and instructed to lay golden eggs. Creativity is never directly sought after. It comes indirectly. It comes while you are complaining about too much routine work, and you decide to spend half an hour on your project. Or while getting ready to lecture, you realize that the textbook is lousy, and that the subject has never been properly explained. While you work at explaining some old material, lo and behold, you get a great new idea. Creativity is

a bad word. Unfortunately, we must leave it in the books because people in power believe in it. It is a dangerous word.

A practitioner's context changes with each step she takes. It is not a fixed thing. We already noted that each practitioner's context is rather unique. To some extent a practitioner's context is a matter of chance, but it is worth considering how to bias that chance.

Her context's initial development mostly happens during her training in a discipline, which tries to lay a groundwork of shared material, a canon, so that two practitioners have enough overlapping context that they can interact. Physics has one of the best developed canons. A student studies classical mechanics, statistical mechanics, electromagnetism, and quantum mechanics. When she enrolls at university, she takes a sequence of courses that go through all of them at a basic level, then a second sequence of courses revisiting them again at a higher level. If she continues to graduate school, she goes through yet another sequence of the four courses, in yet more depth, and then studies the material by herself to prepare for a comprehensive examination. After the examination, she is turned loose in a research lab. The physics canon is very effective at building a common language, though four passes through it delays students becoming practitioners for a long time.

Having been a student, a practitioner can very easily become one again as a means of working on her context. It can be the right approach, but in limited ways. In particular, there is usually some point in a practitioner's career when a purely technical issue is holding her back. Perhaps she needs to be much more competent with microscopes, or lacks a mathematical technique that suddenly has become central to the theories in her discipline. Such cases shouldn't be treated as research problems. Find an expert, a course, or a good exposition. Limit the scope of the material, and the time to be devoted to it. And actively work with the material. For example, I regularly directed colleagues who were planning to do lots of work with microscopes to the Analytical and Quantitative Light Microscopy course at Woods Hole Marine Biological Laboratory, which in ten intense days takes its students through hours of hands on time at a huge range of microscopes, and covers essentially all of modern light microscopy.

Being a student is an obvious way to work on your contexts, but it only works in areas that are already known, which doesn't help a practitioner with her primary work. So how do you develop your context in unknown areas? The first major way is by exposing yourself to the very different contexts of other people, whether practitioners, students, or laymen. Departments and institutes run series of colloquia and seminars purely for the purpose of exposing their members to a regular stream of contexts different from their own. Another way of exposing yourself to different contexts is to explain something to a student or a layman. This often lays bare problems in your understanding. Feynman put it[21]

> The questions of the students are often the source of new research. They often ask profound questions that I've thought about at times and then given up on, so to speak, for a while. It wouldn't do me any harm to think about them again and see if I can go any further now. The students may not be able to see the thing I want to answer, or the subtleties I want to think about, but they remind me of a problem by asking questions in the neighborhood of that problem. It's not so easy to remind yourself of these things.

The other major way to develop your context is to systematically challenge yourself with results. Gian-Carlo Rota recounts:

> Richard Feynman was fond of giving the following advice on how to be a genius. You have to keep a dozen of your favorite problems constantly present in your mind, although by and large they will be in a dormant state. Every time you hear or read a new trick or a new result, test it against each of your twelve problems to see whether it helps. Every once in a while there will be a hit, and people say: "How did he do it? He must be a genius!"[59]

This works the other way as well. In microbiology, I kept a mental collection of various odd organisms in my head that I could use as counterexamples or as the systems for thought experiments. For example, would a notion about vision still work if it were applied in octopi, which see polarization but very few colors? What happens to scaling laws relating metabolic rate to organism size when extrapolated to enormous bacteria such *Thiomargarita namibiensis*, which is hundreds of times larger than the species usually studied in a laboratory?[62] Similarly, an historian will have a huge amount of information about the historical record on tap. A materials scientist will have the properties of a wide range of solids organized in her mind. When such counterexamples throw up a mismatch, there is often something interesting to be examined.

What techniques a practitioner uses to dredge through her foundational context vary widely, but it is important to have a regular regimen so that there is something to draw on when you reach a blank in your planning context.

We have taken the backwards looking model of science developed in the previous chapters and inverted it to understand the view of a practitioner looking forward. We have some concrete notion of what her activities are. Much of it is the day to day labor of running trials or extending theories. That labor is punctuated by examining what gambits are available to her and making short term plans for the next round of labor. And she regularly exposes herself to other people's contexts and to material that challenges her own context in order to dredge material up from her subsconsious foundational context and bring it into her planning.

9 | The problems of interdisciplinary research

Once they have gotten past being a student, the other transition that often trips up practitioners is working on projects that span multiple disciplines. Practitioners of different disciplines have very different foundational contexts, and that difference erects a language barrier between them. For most historians, an organic chemist's talk of aldehydes and electron pushing is so much nonsense, and most chemists are at best hazy on what exactly constitutes historiography. These language barriers are exhausting to fight through.

Unfortunately, open lack of understanding between disciplines is the best case. Practitioners of different disciplines may not realize that there is a language barrier between them, which is more pernicious. To see how this can happen, consider two people standing in front of the Eiffel Tower. They can use their words rather imprecisely when discussing it, since the object itself is before both of them. But two people speaking on the telephone, with one standing in front of the Eiffel Tower and the other standing in front of the Space Needle, may think they are discussing the same object if they are equally vague. Just so, practitioners of the same discipline share much of their foundational context. They can point and wave vaguely, and usually succeed in communicating with each other, since the entities of their discipline are before both of them. But for practitioners of different disciplines, each will try to interpret the vague expressions of the other in terms of her own foundational context.

In more established interdisciplinary areas such as biochemistry these language barriers are not a problem. All biochemists are expected to be conversant with both biology and chemistry, so there are no language barriers to be surmounted. But in less established interdisciplinary areas, a project usually takes the form of putting people from multiple fields together in a room and trying to get them to do something together. Often one side wants some of the technical tools of the other side. While I was in biology there was a fad for trying to wed biology and various mathematical disciplines, which usually meant a biologist getting a physicist or other practitioner to build a mathematical model or some kind of software for them. In such cases the invisible language barrier is ubiquitous. I have many times sat in meetings where I was the only person present conversant in both disciplines represented and watched people run afoul of it.

Given these problems, why bother with interdisciplinary research? Very simply, richness. In the last chapter we said that practitioners unstick their planning by challenging their context. Trying to meld the contexts associated with different disciplines is a dramatic challenge to each one of them. For example, various bacterial species grow at drastically different rates. Some species such as *Escherichia coli* can divide once every twenty minutes, given adequate nutrition. Others, fed to repletion, only divide once a day or even more slowly. Yet often these species occupy similar environments. So do the more slowly growing bacteria respond more slowly to changes in those environments? When posed this question, practitioners studying bacteria whose original training was in biology assume that the slowly growing ones still respond quickly, since they must deal with the same changes in the environment as their fast growing neighbors. Practitioners whose original training was in physics assume that the slowly growing ones respond more slowly, since changing their complement of molecular machinery is limited by how fast they grow and divide unless they are actively destroying existing machinery to make room for more, which is extremely energetically wasteful. Each context challenges the other and raises a very interesting question.

So interdisciplinary research is a useful path forward. But how do you make it work? The ideal case is that everyone involved should be a competent practitioner of each discipline involved. But how do we produce such practitioners? For established interdisciplinary fields such as biochemistry, there are training programs already established. But much of the reason to work at the boundary between disciplines is to get the quick payoff of new ideas that such a boundary can give you. In an established interdisciplinary field, those easy ideas have already been taken, so we really want to know how to do interdisciplinary research in the case where there is no established field, nor an established system of training producing practitioners competent in the fields involved. So how does a practitioner make this transition in the absence of an established system of training?

Most practitioners simply will not. They have already spent years becoming competent in one field. The idea of spending years starting again in another field is daunting; their professional acquaintance and situation is centered in their field; and they may need to use their competence to make a living rather than going back to the beginning in another field. So what are the conditions necessary for a practitioner to learn a new field?

First, there are the issues mentioned above. The practitioner needs the logistical support to start over. That means that she has time to focus on the new field, income to support herself, and won't set her career back by years. That last condition means that the easiest times to make such a transition are when going into graduate school, or as a fairly senior, tenured professor with no research group. In the early years of graduate school, all the students are trying to make the transition from student to practitioner. However, the habits of being a student don't transfer cleanly between fields, so the newcomer will need less time to unlearn being a student, and can spend that time learning a subset of the new field. However, this requires a graduate program that does not impose comprehensive examinations in the subject matter. A senior professor with no research group has enough breathing room to do the same thing.

Next, the practitioner has to have some reason to change fields. My transition from physics to biology happened because, given the graduate schools I got into, I had a choice of living in poverty during graduate school in physics, or being paid comfortably in biology. I transitioned into biology for money. Perhaps others who have made the transition did so for more intellectually pure reasons.

Finally, a practitioner's contacts are all in her original field. Their interests and understanding are in that field, and the social pressures they impose are to stay in that field. So ideally a practitioner transitioning into a new field should be immersed in her new field and isolated from her old one. This is *very* uncomfortable at first. Indeed, it is so uncomfortable that a practitioner trying to to make the transition by herself is likely to retreat. The most successful case is when there is a "beach head" of practitioners who have already made the transition.

We can find examples of both what to do and what not to do to support practitioners transitioning between fields at the Center for Studies in Physics and Biology at the Rockefeller University. In the 1950's and 1960's it was an example of how not to support such transitions. Before that time, the Rockefeller University had been solely a biological research institute. Then it hired a number of professors of physics, mathematics, philosophy, and other fields. These professors carried on working with their former colleagues, training graduate students in their own disciplines, and largely ignoring the biologists else-where on the small campus. There were still a few doing that when I was there in the first decade of the 21st century. It is a measure of this isolation that, when someone gave a lecture at the Rockefeller University going over Marc Kac's classic paper *Can one hear the shape of a drum?*,[40] the biologists who heard it were enthused, and nearly none of them knew that Marc Kac, a practitioner universally known in mathematics, had been on the faculty there until 1981.

On the other hand, when I was there, a thriving school had arisen that carrid practitioners from physics, mathematics, and engineering into biology. The social structure was centered around a pair of theoretical physicists, Eric Siggia and Marcello Magnasco, who were gregarious and interested in everything. Both had one or two students and postdoctoral researchers work-ing directly with them, but most of their students were shared with one of a large number of experimental biology laborato-ries, and spent their time in those laboratories. Both would talk practitioners making the transition into biology through the in-tellectual difficulties. At the same time, Stanislas Leibler, an experimental physicist of broad interests, ran a large laboratory working on biological problems that caught his fancy. His lab contributed another bulk of people making a transition between fields. There were enough such practitioners that the biologists on campus had become accustomed to the issues their new col-leagues were facing, and the "physicists," for so all praciction-ers whose background was not biology were called, had enough people to run a seminar that acted as a meeting point for all involved.

The Rockefeller University also had a few peculiarities of orga-nization which made this possible. All graduate students were, and still are, funded directly by the university, not by the pro-fessors they worked with, which made biologists more willing to take on practitioners transitioning from other fields. And the graduate program, very unusually, accepted applicants with no background whatsoever in biology, and did not require a comprehensive examination of the students' knowledge of biology, which would have deterred most people from trying the transition.

So once you have carved out time and support to transition to a new field, and some force has directed you into a particular field, how do you actually make the transition? The first step is simply to start. Begin with a topic of interest you. Find a piece of primary literature on it. Don't start with a textbook or a review paper. In the primary literature you will find yourself faced with passages like[1]

> We previously suggested that *E. coli mazEF*-mediated cell death is a population phenomenon. Here we confirm that *E. coli mazEF*-mediated cell death was dependent on the density of the bacterial population. Adding rifampicin for a short period to inhibit transcription led to *mazEF*-mediated cell death at densities of 3×10^8 or 3×10^7 cells/ml, but not at 3×10^5 or 3×10^4 cells/ml. Consequently, we examined whether the supernatant of a dense culture could restore *mazEF*-mediated cell death in a diluted culture...[45]

Unless you have some background in molecular biology, this is likely to be incomprehensible. For readers who are molecular biologists, here's a different example:

> The best-known distribution with optimal localization, the Wigner-Ville distribution, achieves optimal

---

[1]The paper this comes from was chosen at random from my citation manager.

localization at the expense of infinitely long range in both frequency and time. Because it is bilinear, the Wigner transform of a sum of signals causes the signals to interfere or beat, no matter how far apart they are in frequency or time, seriously damaging the resolution of the transform. This nonlocality makes it unusable in practice and led to the development of Cohen's class. In contrast, it is readily seen from Eq. 1 that the instantaneous time-frequency reassignment cannot cause a sum of signals to interfere when they are further apart than a Fourier uncertainty ellipsoid; therefore, it can resolve signals as long as they are further apart than the Fourier uncertainty ellipsoid, which is the optimal case. Thus, reassignment with instantaneous time-frequency estimates has optimal precision (unlimited) and optimal resolution (strict equality in the uncertainty relation).[28]

What do you do with passages like this? At this point you need a practitioner of the field that you can talk to. If you can find a practitioner who has made the same transition that you are attempting, this will be even easier, because she will have some idea of what you need to know. But what do you ask about?

We will use the model of a science that we put together earlier in this book as a checklist. First, understand the trials. What precisely are the authors of the passages above describing? The sentence "Adding rifampicin for a short period to inhibit transcription led to $mazEF$-mediated cell death at densities of 3×10^8 or 3×10^7 cells/ml, but not at 3×10^5 or 3×10^4 cells/ml." from the first passage means that the practitioner took several flasks containing bacterial medium, inoculated them with a culture of *Escherichia coli* from her freezer, then let them grow for some time. From time to time she used a pipette to take a little bit of culture and put it in an instrument that measures how much light is absorbed by liquid.

We have to pause here and understand why she would do that. This is another trial. For *E. coli*, light absorption is proportional to the number of cells in the liquid until the culture is extremely dense. She had previously calibrated the absorbence to the number of colonies cultures of different absorbences produced on Petri dishes of agar.

Returning to the original trial, when one of her cultures reached a density of 3×10^4 cells per milliliter, she added the antibiotic rifampicin at some dose described elsewhere in the paper, and allowed it to sit for "a short period" (elsewhere in the paper it says 10 minutes). Then she put cells from the culture in a centrifuge to pellet them at the bottom of a tube, poured off the medium that contained the rifampicin, added some clean medium, shook the cells back into suspension, and spread the liquid out on agar. Once the colonies had grown, she counted how many there were. Then she did the same thing when a culture reached 3×10^5 cells per milliliter, and again two more times when cultures reached 3×10^7 and 3×10^8 cells per milliliter.

Figuring this out without a current practitioner to explain it would be nearly impossible. The next stages would be even harder. What is the ideal trial behind this trial? What primitive notion is it measuring a value of? What makes this trial's execution reasonable? The idealized trial is performing some operation on cell cultures to produce measurements of what fraction of cells survive from cultures with different densities.

There are lots of details of what makes a trial reasonable. The best way to start internalizing them is to try to design similar trials, and run them past the practitioner that you are working with. That is also the next obvious step to begin to internalize the idealized trials and the gambits that lead to them. Ideally, you would go into a lab and run some of these trials that you have designed.

Once you have understood the trials in the paper, the idealized trials that they are approximating, and what you need to look for to check that they are reasonable, then you must look for the theories that made up the context that generated them. For example, in the quorum sensing trial above, you would need to learn about toxin-antitoxin complexes, which are pairs of proteins once of which, the toxin, is lethal to the cell producing it, the other of which, the antitoxin, binds the first and prevents it from acting. Under particular triggers, the antitoxin ceases binding the toxin, and the cell suicides. At the period when this paper was writen, there was a fad for explaining phenomena via toxin-antitoxin complexes. You would need to know that quorum sensing in most systems where it has been detected occurs via cells secreting some small molecule, so one of the theoretical patterns of the field is to look for those molecules. You would need to know that the antibiotic rifampicin inhibits transcription, which prevents the bacteria from producing new proteins after the antibiotic affects them, so mechanisms they see must be due to preexisting proteins, such as toxin-antitoxin complexes. If you are deeply into antibiotics, you would also need to know that the standard theory of rifampicin's action—inhibiting the transcription of DNA into RNA to make protein—seems to apply to the vast majority of kinds of RNA produced, but not to all of them. And you would need yet more. The context of a real practitioner is enormous.

And, having dug through all of this, you pick up another piece of primary literature and do it again.

In the course of all this, you are likely to have dug through a large number of other pieces of primary literature and spent hours in discussion with the practitioner who is guiding you. Then, having gotten through this one paper, you do it again with another one. The good news is that the second paper in an area will go faster than the first. After a while you start to have seen the idealized trials that practitioners use in your new field, and this work goes faster. Eventually your context is complete enough where the short term plans you can generate for yourself involve trials and theoretical work, not trying to understand what drove someone else's actions.

It should be obvious that this process is slow and frustrating. It may be worse since some fields will have significant technical prerequisites, such as mathematics to do physics, musical training for musicology, or a knowledge of Chinese to study the history of China. To quite Halmos, "The beginner...should not be discouraged if...he does not have the prerequisites to read the prerequisites."[32]

Given all this, it's important to set your expectations. You will likely spend the first six months of immersion in the new field not understanding what people are saying to you, and then the next six months will be spent unable to make yourself understood. This is perfectly normal. And that first year, after which you can competently communicate, may be enough. A team of researchers from different disciplines can work together on a project if there are at least a few people involved who can bridge the fields. A team of molecular biologists, condensed matter physicists, and psychologists can work together if there is a practitioner involved who knows biology and condensed matter physics, another who knows biology and psychology, and one more who knows psychology and condensed matter physics; or a single practitioner who knows all three. This kind of collaboration is by far the fastest way to produce interdisciplinary research. Unfortunately, all too many institutions and teams try to skip the painful year or so of retraining that individuals involved need and simply throw together groups of researchers. Without individuals to act as bridges, language barriers all too easily sabotage the whole group's work.

You will also face large amounts of mental dissonance. During your training in your first discipline, you internalized notions of "understanding" and "interesting." One of the biggest changes that learning a second discipline makes in a practitioner is dredging those notions into her conscious mind. They must be pulled up and examined. They will not apply in the new discipline, but they will cause feelings of revulsion at the new notions that you are trying to internalize. Fortunately, having done this once, you don't have to do it again. Learning a third or a fourth discipline does not involve the same emotional turmoil.

Interdisciplinary research gets bandied about as a panacea at various times, but rarely does anyone talk about what actually has to happen in order for it to work. Institutions must provide the support for transitions. If an institution wants practitioners to transition to a particular field from other fields, it must provide a path for practitioners to do so. That means financial support, a way of immersing practitioners in the new discipline, removing burdens like comprehensive examinations, and ensuring that the transition does not sabotage the practitioner's career prospects. It means providing practitioners of the target discipline with time and social structures in which to guide those making the transition into the discipline. And ideally it means forming a critical mass of those making the transition in order to make the process easier for those involved.

For individuals, it is a deep change. It takes aspects of being a practitioner that are usually unconscious and drags them into the conscious mind. It takes them out of the environment where they had established themselves. And it makes it uncomfortable for them ever to return. Thereafter they are neither fish nor fowl. The recompense is the rich vein of material to found at the boundaries between disciplines.

Bibliography

[1] Ahn, C.P. et al. 2014. The tenth data release of the sloan digital sky survey: First spectroscopic data from the SDSS-III apache point observatory galactic evolution experiment. *The Astrophysical Journal Supplement Series*. 211, 2 (2014), 17.

[2] Altman, J. and Das, G.D. 1967. Postnatal neurogenesis in the guinea-pig. *Nature*. 214, (1967), 1098–1101.

[3] Bell, J.S. 2011. *Speakable and unspeakable in quantum mechanics*. Cambridge University Press.

[4] Beringer, J. et al. 2012. Review of particle physics. *Physical Review D - Particles, Fields, Gravitation and Cosmology*. 86, 1 (2012), Chapter 1, Figure 2.

[5] Bhatt, A. 2010. Evolution of clinical research: A history before and beyond james lind. *Perspect Clin Res*. 1, 1 (Jan-Mar 2010), 6–10.

[6] Biederman, I. and Shiffrar, M.M. 1987. Sexing day-old chicks: A case study and expert systems analysis of a difficult perceptual-learning task. *Journal of Experimental Psychology: Learning Memory, and Cognition*. 13, 4 (1987), 640–645.

[7] Blane, G. 1815. Statements of the Comparative Health of the British Navy, from the year 1779 to the year 1814, with proposals for its farther improvement. *Medico-chirurgical transactions*. Royal Society of Medicine Press.

[8] Bloor, D. 1976. *Knowledge and social imagery*. Routledge & Kegan Paul.

[9] Bolhuis, H. et al. 2004. Isolation and cultivation of Walsby's square archaeon. *Environmental Microbiology*. 6, 12 (Dec. 2004), 1287–1291.

[10] Borges, J.L. 1962. Pierre Menard, Author of the Quixote. *Labyrinths*. New Directions.

[11] Borovik, A.V. 2009. *Mathematics under the microscope*. American Mathematical Society. pp.186.

[12] Brent, L. in Nature: Animal Odd Couples (Season 31; November 7, 2012), 16:00-17:00.

[13] Center for Applications of Psychological Type September 12, 2015. The story of Isabel Briggs Myers. *http://www.capt. org/mbti-assessment/isabel-myers.htm*.

[14] Daston, L.J. and Galison, P. 2010. *Objectivity*. Zone Books.

[15] Diamond, J. 1997. *Guns, germs, and steel*. W. W. Norton.

[16] Durant, W. and Durant, A. 1935–1975. *The story of civilization*. Simon; Schuster.

[17] Eddington, A.S. 1928. *The nature of the physical world*. Cambridge: Cambridge University Press. pp.74.

[18] Ehret, C. 2001. *An African Classical Age: Eastern and Southern Africa in World History 1000 BC to AD 400*. University of Virginia Press.

[19] Elowitz et al. 2002. Stochastic gene expression in a single cell. *Science*. 297, 5584 (Aug 16 2002), 1183–1186, PMID: 12183631.

[20] Feynman, R.P. 1985. Cargo cult science. *Surely You're Joking, Mr. Feynman!* W.W. Norton & Company. pp.308–318.

[21] Feynman, R.P. 1985. The dignified professor. *Surely You're Joking, Mr. Feynman!* W. W. Norton & Company. pp.149–150.

[22] Feynman, R.P. 1965. *The Feynman Lectures on Physics, Volume 1*. Addison Wesley. pp.26–3.

[23] Fischer, D.H. 1970. *Historians' Fallacies: Toward a logic of historical thought.* Harper Perennial. pp.25.

[24] Fisher, R.A. 1926. *Statistical methods for research workers.* Edinburgh/London: Oliver & Boyd.

[25] Floderus, B. et al. 1988. Smoking and mortality: A 21-year follow-up based on the swedish twin registry. *International Journal of Epidemiology.* 17, 2 (1988), 332–340.

[26] Friedman, S.G. 2005. Straight talk about parrot behavior. *Good Bird Magazine.* 1, 3 (2005), 18–20, 62–68.

[27] Gal, O. and Chen-Morris, R. 2013. *Baroque science.* University of Chicago Press.

[28] Gardner, T.J. and Magnasco, M.O. 2006. Sparse time-frequency representations. *Proceedings of the National Academy of Sciences.* 103, 16 (Apr. 2006), 6094–6099.

[29] Gintis, H. 2009. *Game theory evolving.* Princeton University Press.

[30] Greenspan, S.I. and Lewis, N. 2009. *Building healthy minds: The six experiences that create intelligence and emotional growth in babies and young children.* Da Capo Press. pp.215–221 (Kindle locations).

[31] Hacking, I. 1988. Telepathy: Origins of randomization in experimental design. *Isis.* 79, 3 (1988), 427–451.

[32] Halmos, P. 1950. *Measure theory.* Litton Educational Publishing, Inc. pp.(v).

[33] Hamming, R.W. 2005. *The Art of Doing Science and Engineering.* Taylor & Francis e-Library. pp.189–190.

[34] Hamming, R.W. 2005. *The Art of Doing Science and Engineering.* Taylor & Francis e-Library. pp.186.

[35] Heisenberg, W. 1925. Quantum-theoretical re-interpretations of kinematic and mechanical relations. *Zs. Phys.* 33, (1925), 879–893.

[36] Henrici, A.T. 1928. *Morphologic variation and the rate of growth of bacteria.* Charles C. Thomas.

[37] Hermann International September 12, 2015. Origins of the Whole Brain® Thinking System and HBDI® Assessment. *http://www.herrmannsolutions.com/what-is-whole-brain-thinking-2/*.

[38] International Enneagram Association September 12, 2015. History of the Enneagram. *http://www.internationalenneagram.org/enneagram_history/index.html*.

[39] Judkins, G. et al. 2008. Determinism within human-environment research and the rediscovery of environmental causation. *The Geographical Journal*. 174, 1 (March 2008), 17–29.

[40] Kac, M. and Weinberger, H.F. 1974. Can You Hear the Shape of a Drum? *The American Mathematical Monthly*. 81, 5 (May 1974), 534.

[41] Kampen, N. van 1988. Ten theorems about quantum mechanical measurements. *Physica A: Statistical Mechanics and its Applications*. 153, 1 (1988), 97–113.

[42] Kaprio, J. and Koskenvuo, M. 1989. Twins, smoking and mortality: A 12-year prospective study of smoking-discordant twin pairs. *Social Science & Medicine*. 29, 9 (1989), 1083–1089.

[43] Kirksey, M.A. et al. 2011. Spontaneous loss of phthiocerol dimycocerosate (pDIM) production in mycobacterium tuberculosis h37Rv. *Infect Immun*. 79, 7 (2011), 2829–2838.

[44] Koch, R. 1882. Die aetiologie der tuberkulose. *Berliner Klinische Wochenschrift*. 15, 221–230 (1882).

[45] Kolodkin-Gal, I. et al. 2007. A Linear Pentapeptide Is a Quorum-Sensing Factor Required for mazEF-Mediated Cell Death in Escherichia coli. *Science*. 318, 5850 (Oct. 2007), 652–655.

[46] Kroenke, K. et al. 2001. The PHQ-9: Validity of a brief depression severity measure. *J Gen Intern Med*. 16, 9 (September 2001), 606–613.

[47] Lenski, R. September 19, 2014. Summary data from the long-term evolution experiment. $http://myxo.css.msu.edu/ecoli/summdata.html$.

[48] Lorenzi, R. September 20, 2013. Discovery News. Ancient Etruscan prince emerges from tomb. $http://news.discovery.com/history/archaeology/ancient-etruscan-prince-tomb-130920.htm$.

[49] Lu, Y. and Cross, F.R. 2010. Periodic cyclin-Cdk activity entrains an autonomous Cdc14 release oscillator. *Cell.* 141, (April 16 2010), 268–279.

[50] McLeod, S. 2011. Warriors and women: The sex ratio of Norse migrants to eastern England up to 900 CE. *Early Medieval Europe.* 19, 3 (2011).

[51] Myneni, K. December 11, 2014. Symmetry destroyed: The failure of parity. $http://ccreweb.org/documents/parity/parity.html$.

[52] Otto, T. 2009. What happened to cargo cults? Material religions in Melanesia and the West. *Social Analysis.* 53, Issue 1 (Spring 2009), 82–102.

[53] Pachter, L. February 10, 2014. The network nonsense of Albert-László Barabási. $https://liorpachter.wordpress.com/2014/02/10/the-network-nonsense-of-albert-laszlo-barabasi/$.

[54] Peirce, C.S. and Jastrow, J. 1885. On small differences of sensation. *Memoirs of the National Academy of Sciences.* 3, (1885), 75–83.

[55] Planck, M. 1968. Scientific autobiography and other papers. Williams & Norgate Ltd. pp.33–34.

[56] Rende, R.D. et al. 1990. Who discovered the twin method? *Behavior Genetics.* 20, 2 (1990), 277–285.

[57] Ross, S.A. personal communication.

[58] Rota, G.-C. 2008. A mathematician's gossip. *Indiscrete thoughts.* Birkhäuser. pp.210.

[59] Rota, G.-C. 2008. Ten lessons I wish I had been taught. *Indiscrete thoughts*. Birkhäuser. pp.202.

[60] Sachs, J.L. and Bull, J.J. 2005. Experimental evolution of conflict mediation between genomes. *PNAS.* 102, 2 (Jan 2005), 390–395.

[61] Scerri, E.R. 2006. *The Periodic Table: Its Story and Its Significance.* Oxford University Press. pp.29–100.

[62] Schulz, H. et al. 1999. Dense populations of a giant sulfur bacterium in namibian shelf sediments. *Science.* 284, 5413 (April 16 1999), 493–495.

[63] Shapin, S. and Schaffer, S. 2011. *Leviathan and the air-pump: Hobbes, boyle, and the experimental life.* Princeton University Press.

[64] Simanek, D.E., 2012. A short history of the search for perpetual motion. *http://www.lhup.edu/~dsimanek/museum/peopl* *people.htm.*

[65] Swain et al. 2002. Intrinsic and extrinsic contributions to stochasticity in gene expression. *Proc Natl Acad Sci USA.* 99, 20 (Oct 1 2002), 12795–12800, PMID: 12237400.

[66] Tischler, A.D. et al. 2012. Mycobacterium tuberculosis requires phosphate-responsive gene regulation to resist host immunity. *Infection and Immunity.* 81, 1 (Dec. 2012), 317–328.

[67] Toynbee, A.J. 1935–1961. *A Study of History.* Oxford University Press.

[68] United States Census Bureau December 21, 2010. 2010 Census. *http://www.census.gov/2010census/.*

[69] Weinert, T.A. and Hartwell, L.H. 1988. The RAD9 gene controls the cell cycle response to DNA damage in Saccharomyces cerevisiae. *Science.* 241, (July 1988), 317–322.

[70] Weingarten, J. October 6, 2013. How a prince became a princess. *http://judithweingarten.blogspot.com/2013/10/how-prince-became-princess.html.*

[71] Wick, L.M. et al. 2002. The apparent clock-like evolution of Escherichia coli in glucose-limited chemostats is reproducible at large but not at small population sizes and can be explained with Monod kinetics. *Microbiology.* 148, 9 (2002), 2889–2902.

[72] Woese, C.R. and Fox, G.E. 1977. Phylogenetic structure of the prokaryotic domain: The primary kingdoms. *Proc. Natl. Acad. Sci. USA.* 74, 11 (November 1977), 5088–5090.

[73] Vaughan v. Menlove, (1837) 3 Bing. N.C. 467, 132 E.R. 490 (C.P.).

About the author

Frederick J. Ross received a classical education from his parents while growing up in the wilds of Virginia, then worked in physics, mathematics, microbiology, and statistics in the US and Europe before leaving academia. He has played violin for twenty some years, written some fiction, and apparently annoyed a lot of people. Now he lives on an island in the Salish Sea with his wife and children.

You can find out about other books, or read essays and commentary by Frederick J. Ross at `http://madhadron.com`.